cracked media

The MIT Press Cambridge, Massachusetts London, England

cracked media: the sound of **malfunction**

caleb kelly

MIT Press books may be purchased at special quantity discounts for business or sales promotional use. For information, please email special_sales@mitpress.mit.edu or write to Special Sales Department, The MIT Press, 55 Hayward Street, Cambridge, MA 02142.

This book was set in Helvetica Neue and Nikona by Graphic Composition Inc., Bogart, Georgia. Printed and bound in the United States of America.

Library of Congress Cataloging-in-Publication Data

Kelly, Caleb, 1972– .

Cracked media : the sound of malfunction / Caleb Kelly.

p. cm.

Includes bibliographical references (p.) and index.

ISBN 978-0-262-01314-7 (hardcover : alk. paper)

1. Avant-garde (Music). I. Title.

ML197.K495 2009

786.7—dc22

2008044263

10 9 8 7 6 5 4 3 2 1

contents

acknowledgments

The concept for *Cracked Media* initially came to me while listening to a track from the German band Oval. It was 1997 and the music was so foreign to me at that time that I couldn't help but try to grasp the reason for its existence. Some years later this problem lead me to Sydney where I was supervised in doctoral research by Douglas Kahn. He was certainly a exacting taskmaster, which led to long one-on-one meetings and many, many failed attempts at getting words on paper. I wish to thank him for his continuing support. After Doug returned to America I changed supervisors for a recent graduate of Doug's, Mitchell Whitelaw. Mitchell had the good luck of getting the final writing and rewriting stages of the research, and I am heavily indebted to his care and attention to words and their stringing together. Both Doug and Mitchell are key figures behind this research and I thank them again here.

I am extremely grateful and lucky to have had the advice and input from two primary readers, Peter Blamey and Ann Finegan. These two have at varying points in the writing of this book given advice and positive support. I also feel great gratitude to Roger Conover for his support from my initial queries through to the publication of the book and to Judith Feldmann for her meticulous editing of the text.

My thanks is also extended to all those who played a part in this research over the years, many through various conversations, formal and informal. Reiko Tomii has aided in my attempts to understand issues in Japanese culture and seems always available for advice. I would specifically like to thank these practitioners (I still haven't worked out a word that covers the art–music–performance–composition divide) who have been especially generous with their time: Nicolas Collins, Michael Graeve, Milan Knížák, Christian Marclay, Frank Metzger, Sebastian Oschatz, Philip Samartzis, and Yasunao Tone.

Finally, for reading and rereading, financially supporting me, and staying on my case, I wish to thank Lynn Kelly.

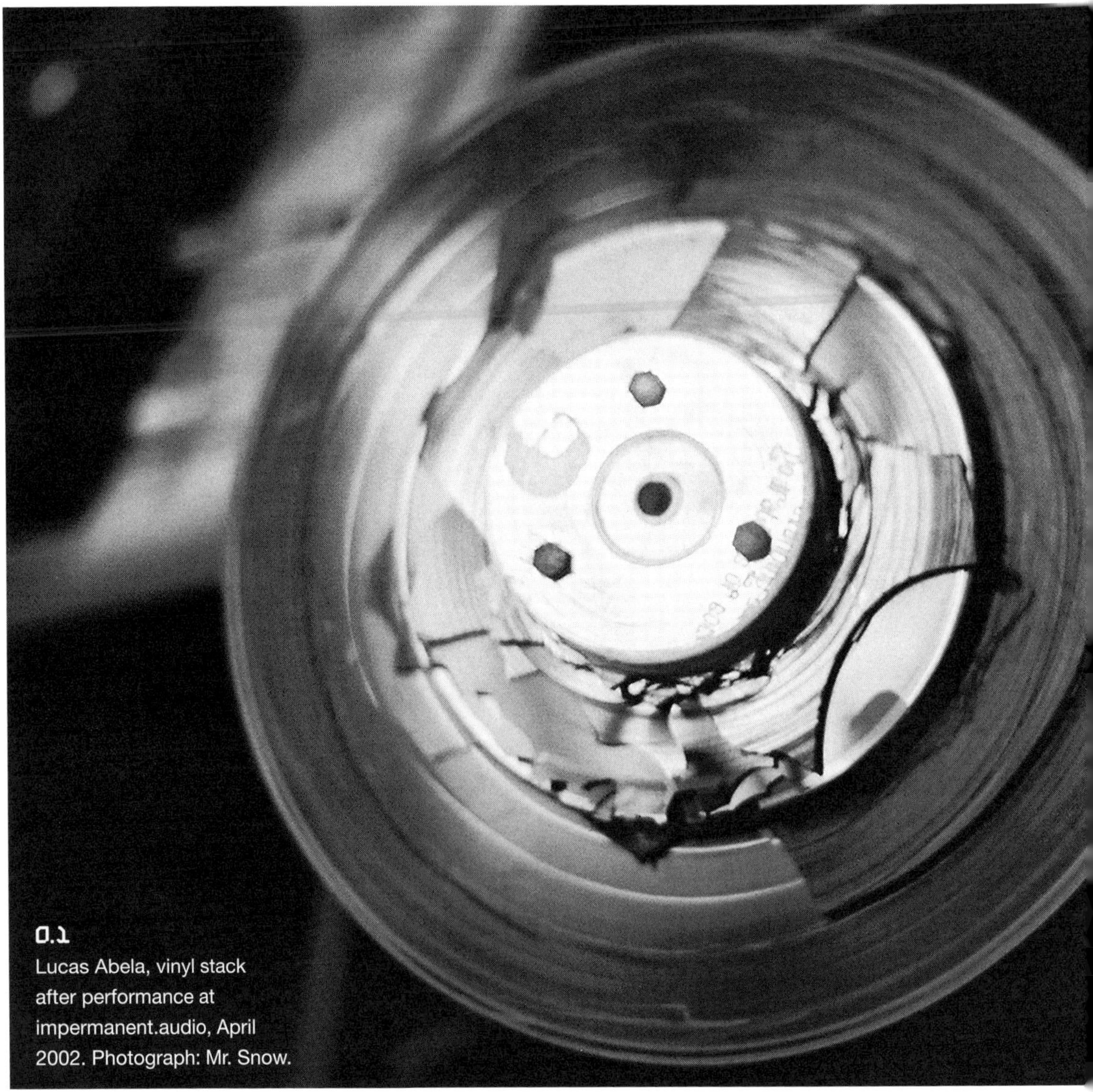

0.1

Lucas Abela, vinyl stack after performance at impermanent.audio, April 2002. Photograph: Mr. Snow.

introduction

In a small warehouse gallery space in the center of Sydney a noise turntablist, Lucas Abela, stands over a stack of vinyl that is bolted to the rotary device of an industrial sewing machine. Instead of spinning the records at 33 or 45 revolutions per minute this "turntable" spins the records at up to 2,850 revolutions per minute. The stylus itself is made of a meat skewer directly connected to a guitar amplifier. Abela holds the makeshift needle against the records with difficulty; the speed of their rotation constantly whips his stylus off the vinyl discs. The sound is dominated by intense noise: loud bursts of sound join occasionally recognizable high-speed cartoon-like tunes. The force with which the needle is applied to the discs causes them to shatter at regular intervals, and as they do shards of vinyl are propelled across the room; some are firmly embedded in the gallery walls. In total the performance lasts around eight minutes before the combination of a completely shattered set of vinyl and a broken needle causes the end of the piece.[1]

In the industrial architecture of the Brisbane Powerhouse in Australia, Michael Graeve crouches around an array of old turntables on the floor of a raised concrete block in the center of the old Powerhouse building. Graeve is playing as part of the What is Music? festival in February 2004. These old turntables are spun without a record in sight; instead the cartridges knock, bounce, and thump around their platters—some with felt covers, some without. A strong rasping sound emanates from these antique technologies, and the audience, members of whom join Graeve on the stage, watch and listen to the developing noisy drone.

Vicky Browne, a sound artist from New Zealand, has installed an exhibition at Pelt gallery in Sydney entitled *Time Table* (2006). At the center of the exhibition are four turntable works, *Sticks and Stones*, *Old and New*, *Night Music*, and *Morning Music*. The first work is a "record disc" made of twigs; the second a compact disc of Browne singing, which is played on a turntable with a turntable stylus; the third is an aluminum record of bird calls; and the fourth a copper record of bird calls. The effect of placing a stylus on these foreign objects in 2006 effects a slippage between the nostalgia for playing records and the strangeness of playing objects that are not usually thought of as records.

Throughout the late twentieth century and into the twenty-first, it became almost common to see performances that used some element of the manipulation, breaking, or destruction of sound mediation technologies in live performance. The image of the performer bent over newly released audio technology, or alternatively, over beaten up domestic hi-fi equipment, producing an array of noises at extreme volume or at palpable near silence, was a regular part of many experimental music performances. The performance of these activities was not geographically defined; they could be witnessed across North America, throughout Europe, from Australia to New Zealand, and in Japan. The events themselves were held in a multiplicity of venues: from pristine concert halls to dirty rock pubs; from white-walled art galleries to illegal warehouse squats; from highly attended festivals to unpublicized performances in front of a few devotees. The music itself was as varied as the countries in which it was performed. At times it was tremendously excessive and loud, literally causing the paint on the ceiling to crumble onto the audience; at other times it was so quiet the audience's breath could be heard. The music also took multiple forms in content and intent: abstract and clean, focused and single minded, dirty and brief, cut-up and filled with differing and discrepant sounds.

"Cracked media" are the tools of media playback expanded beyond their original function as a simple playback device for prerecorded sound or image. "The crack" is a point of rupture or a place of chance occurrence, where unique events take place that are ripe for exploitation toward new creative possibilities. As we will come to see, the crack takes a variety of forms, much like the practices introduced above, from gentle coaxing of faint crackle on the surface of a vinyl record to the total destruction of the playback tools. The practice utilizes cracks inherent in the media themselves—we cannot play a vinyl record without causing some damage to the surface of the disc—and leads to a creative practice that drives playback tools into territory where undesired elements of the media become the focus of the practice. For example, the practitioner might engage in the audio created by a damaged and stuttering CD, rather than trying to rectify the perceived problem by buffing the underside of the disc to remove any dust and dirt.

It verges on cliché to suggest that a major part of the creative processes of artists and musicians is to transform and extend already existing practices and modes of practicing. This might take the form of very slight and simple shifts in methods of production, for example, experimenting with different ways to paint, such as with one's fingers or with the tip of a paintbrush handle; or it might take a much more extreme approach, forcing unexpected sounds out of an instrument, such as a piano, by inserting objects into the strings. In both examples, preexisting and expected methods of use are extended as part of a creative process that seeks new ways to use and transform creative tools, as well as looking for novelty and unique approaches to creative practices. The use of the "crack" as a process is an example of such practice; it is this process that unifies the diversity outlined above. The inquisitive artist, on finding a technology that is new to him or her—be it a newly developed tool just released into the market or an outmoded technology found in a dusty corner of the studio—sets out to see how it works and discover the boundaries and limitations of the device. What can this tool do, and how can I use it in a way that may not have been originally intended? This might be achieved by simple manipulation or modification (taking the technology apart and trying to put it back together), or it might be through overloading it or otherwise stretching its operating parameters, until it starts to fall apart or break down. In this process we find new ways of performing or ways of producing new and unique sounds. There is nothing new in this idea: the painter who uses the brush handle on the canvas and the guitarist who plucks the strings around the head of the electric guitar are both engaged in a similar area of practice. Experimentation with readily available tools and resources is central to contemporary artistic practice and is at the heart of the crack. Here we encounter the experimentalist who is prepared to extend his or her instrument to the point at which it breaks, perhaps never again to be used in the manner in which it was intended. This risk of sometimes great loss is turned to great gain as traditional and commonplace sound practices are themselves transformed, extended, and expanded.

Glitching Digital Audio

The *glitch*, the digital tick caused by lost or incorrect binary code, is both the end point and the starting point for this book. It will be argued that the practice identified as "glitch," which became popular in the late twentieth century, was a key marker in the development of digital arts practices. It represented a familiarity with computer-based technologies, as well as a new direction for their use. It will be my conjecture that the glitch had already been signaled in the varied practices of the crack and the break throughout the twentieth century, and it is against this backdrop that the discussion will ensue.

In the mid- to late 1990s experimental music witnessed an outpouring of interest as the tools of music production were transformed and rapidly expanded with the mass take-up of digital technology. The general population gained access to more affordable computer and home studio equipment, and a surge of experimentation took place. This production found its outlet in numerous small, independent record labels, which released short runs of literally thousands of titles. Glitch, as the genre became known, developed as a central initial part of this outburst of creation. Producers took these newly developed, or newly accessible, musical tools and extended their use well beyond what their designers intended, pushing them until they collapsed or simply stopped working. The cliché of the period, roughly 1994–2000,[2] was the image of the bedroom producer hunched over his or her desktop computer trying various methods to overload its central processing unit (CPU) so as to produce a new digital tick, pop, or click that could be sampled and then sequenced for the next track. Rob Young in the English music magazine *The Wire* wrote:

> Crackles, pops, pocks, combustions, gurgles, buzzes, amplitude tautening, power spikes, voltage differentials, colliding pressure fronts, patterings, jump-slices, fax connections, silent interjections, hums, murmurs, switchbacks, clunks, granulations, fragmentations, splinterings, roars and rushes have overwhelmed the soundscape.[3]

This "noising up" of the digital was part of the noisy project of twentieth-century experimental music, as glitch music combined the "clean" world of the digital with a "dirty," detritus-driven sound that switched the ratios of signal to noise in the realm of digital production. A plethora of artists released music in this period: some of the better known include Pimmon, Oval, Pita, Fennesz, Farmers Manual, SND, Carsten Nicolai, Kim Cascone, Hecker, Microstoria, Autopoesis, Kid 606, Gas, Vladislav Delay, Rehberg & Bauer, Oren

Ambarchi, and Pole. The work of these artists, and many more, opened the ears of other producers already working in other genres and subgenres—including electronic dance music, electronica, drum 'n' bass, industrial rock, and pop—to an expanded sound field.

A central technique of glitch is the *crack*, a term that refers to physical breaks in the surface of an object but also nods toward "cracked" software—software that has been modified in order to remove its copy protection and enable "pirate" reproduction. New Zealand improviser Dean Roberts explicitly used the term for his album *All Cracked Medias* (Mille Plateaux, 1998). Roberts stated that the tracks on the album represent an interest in music production that explores sounds that relate to each other, "through the cracks in the media."[4]

Cracked software was at the center of a scene that was forced, often through lack of financial means, to make do with what applications it could obtain for free. Cracked software, although free, is also unreliable and often unpredictable; combined with the technical demands of real-time digital sound production, it created a production environment that was filled with unplanned occurrences. This instability was harnessed by the burgeoning subgenre, which utilized deliberate accidents for the content of its works. This "cracked" approach bears many similarities with earlier cracked media and will be discussed throughout the book.

The movement was international in context and distributed through traditional record sales as well as through new forms of distribution including the internet, MP3s and CD-Rs.[5] For a brief moment, glitch seemed to fulfill Jacques Attali's prediction of a final "phase" of music that would be centered around general access to the tools of musical production and the production of music for oneself rather than as a form of stockpiled capital.[6] However, this explicitly glitch- and error-driven scene rapidly became assimilated by the wider fields of electronic and popular music production, in their perpetual search for sonic novelty. The techniques developed in the late 1990s are now a common part of contemporary production and for the most part have not been the sole focus of experimental music since the early 2000s. These sounds are now simply another part of the sound palette of the digital producer.

The glitch was a catalyst for the conception of *Cracked Media*. There was a high level of excitement around the glitch work being produced: experimental music record labels, such as Mille Plateaux, were starting up all around the world and audiences rapidly grew for live experimental music events. The work had an underground and countercultural feel, while using new high-tech tools. In performance, musicians often played the music on laptops. There was often an overt disjunction between seeing a solo musician standing calmly on stage in front of a laptop and the extreme volume and sheer density of the noise that was produced, especially by those associated with the Viennese label Mego.[7]

For many the experience of hearing seminal glitch artists Oval for the first time was formative. I first heard the band in 1997 at a friend's place. I was listening to New Zealand noise-rock as well as the aggressive dance genre drum 'n' bass at the time. The sound of Oval's glitching CDs set at an ambient pace was mesmerizing and exciting: the strangeness of this beautiful music mixed with the grating nature of the CD skip caused a reevaluation of music for myself and many others. This was simply due to the feeling that many of the musical genres at the time were stagnating into obvious patterns and the continual and rapid development of scene-driven subgenres and sub-subgenres. Oval dispensed with this, producing a new-sounding music and to many providing a new way of approaching the production of music. In the following years many CDs produced with the sound of malfunction were released in what was an exciting and productive period for experimental music. Through my interest in the music being created at the time I came to identify and reflect on the processes being used by these practitioners. Many of the producers working in the area were unaware of the precedents of their own methods, and I started looking into previous producers of experimental musics and their approaches to various media and traditional instruments.

The glitch in electronic experimental music raises a number of questions that are also relevant to this discussion. Is the use of the glitch a creative dead end, or is there a future in this process? How many times can you force something into failure before it becomes creatively uninteresting? These questions will be addressed by looking back at various approaches to cracked media. Here the aim is to better understand these practices through examining their history.

Music and Sound

At first glance many of the sounds discussed in this book are closely associated with music, as opposed to other types of art. For example, John Cage, Yasunao Tone, Oval, Nicolas Collins, and Otomo Yoshihide,[8] all of whom have used cracked media, are all musicians and/or composers. However, although many of the practitioners of the crack and break do maintain a close connection with music, many of these practices are also more closely aligned to other arts practices, especially the visual arts. Christian Marclay, for example, has a visual art background, and practitioners such as Nam June Paik are very much a part of twentieth-century developments in the visual and media arts. Accordingly, the discussion does not seek to support any particular discipline-specific approach to sound, but rather listens closely to the sounds being produced through these often destructive practices. Cracked media slides across disciplines, between music, sound, and noise. It is part of the wider project of sound in the arts, an area that has recently been receiving attention from a range of different disciplines.

Traditionally, Western art music has drawn a significant distinction between musical and nonmusical sound, until its often noisy introduction in the early stages of the twentieth century. Musical sounds were sounds produced from traditional instruments, played in the correct manner by trained musicians. Other sounds, such as noise and sounds from the outside environment, were deemed "extramusical." There was a clear demarcation between sound that could be controlled, and as such was available as musical material, and sound that was not controllable and was therefore outside of music. Douglas Kahn addresses this distinction in the chapter "The Sound of Music" in his book *Noise, Water, Meat: A History of Sound in the Arts.* He argues that "There has been a line drawn between sound and musical sound,"[9] and points out that Western art music concentrated on music of the past centuries and traditional instruments. The rapid expansion of music's sonic palette occurred when musical sound began to be associated with the wider area of general sound.[10] Furthermore, it was not necessarily music that led the way for sound in the arts, as might be expected. Instead, "some of the most provocative artistic instances of sound came from literature and other writings."[11] As we will see later, nonmusical sound was championed by numerous twentieth-century avant-garde composers, not least of whom was Luigi Russolo, who called for the noise of the modern city to enter music. This call was heard in the twentieth century, and music became a much "nosier" art.

Furthermore, a major twentieth-century shift occurred with the introduction of *all* sound into music. Here extramusical sound joined musical sound in an all-inclusive sonic realm. Elsewhere, Kahn points to modernist artistic strategies that went "outside the given confines of musical sound to other sources of sonic plenitude."[12] One source for such plenitude lay not in other types of music but in “sound.” As Kahn writes:

> what was “outside” music was under constant negotiation, but the main sources of these sounds were nature, urbanism, war, the quotidian and domestic, other languages and dialects, the outward expressions of exceptional psychological states, and technology, whether it was the technology of new instrumentation (musical or scientific), communications (especially phonography and radio), or the wider world of machines and motors.[13]

It was the invasion of all possible sounds into music, or in fact into an area that may simply be called sound (be it music like sound, or art like sound, or any other “soundful” area of the arts), that led to the expansion of these areas. More than merely opening music up to greater possibilities, all sound provided a radical openness for music and for sound to be fully heard in other disciplines. It is this broad field of sound that delimits this discussion, rather than specific disciplines such as music or art.

For the purposes of this book I wish to explain my own position on the term "sound art." In recent years there has been debate over the term's correct meaning and use. The issue, it seems to me, stems from confusion over the term itself and a widening of its use to the point of redundancy. The phrasing of the term suggests a direct link between "sound" and "art," and it was originally used to describe a type of art that investigated sound. I would argue that sound art should be understood as clearly being a subset of art, and it is a genre or area of art that engages with issues pertinent to art practice. If this is the case then the term describes a very small area of practice that utilizes sound and thus is not useful when describing the plethora of sound that is to be found in the arts. In fact the term "sound art" would be more useful if it had an "s" added at the end to form "sound arts," as for the most part those who use the term would rather a much wider usage than the visual arts and could then include music, literature, and performance under its coverage.

An issue with current use of "sound art" as a phrase is that it has been expanded to include experimental musics. "Sound art," however, does not suggest music at all, except through a confusion between the link of sound and music. Part of the rationale of using "sound art" to describe experimental music is to wrest the practice away from the traditions of music and thereby allow it various freedoms not normally extended to traditional music. At this point in the history of music I feel it is high time that the term "music" be taken back by contemporary practitioners and used without concern for its recent shaky past in the hands of the academy. Thus I am comfortable using "music" to describe much of the work discussed within this text.

For my own purposes and the purposes of this book, sound art is defined very specifically as art that has a sound focus. It includes the gallery-based work of Nam Jun Paik, Milan Knížák, and Christian Marclay, among others, but it does not include the music and performance-based practices of Otomo Yoshihide, Oval, and Yasunao Tone. Marclay, and to some extent Paik and Tone, call into question this definition, as they have two related but I believe quite different approaches to their practice. In the case of Marclay, there is his gallery-based art practice and his performance-based music practice. The former is included in my understanding of sound art, while the later is understood as music.

The use of modified, cracked, or broken technologies originally intended for the mediation of sound is a key element in the expansion of musical sound, and such devices can be understood as extended and expanded instruments. In seeking new and unique sounds beyond those created within the traditional framework of Western art music, the historical avant-garde used a number of techniques. It invented new instruments and nontraditional sound producing devices, but also used extended techniques and prepared instruments to expand the sound created by traditional instruments. Extended techniques are performance techniques that extend the sound of the instrument by playing it in ways not originally intended, for example, playing a violin with the wooden side of the bow or plucking it with a foreign object. Prepared techniques, by contrast, are techniques whereby the instrument itself is altered in some way, usually impermanently, before a performance.[14] Many of the cracked sounds discussed here fall into these categories. For example, a turntable may be modifed by replacing the stylus with a toothpick, or Scotch tape may be added to the substrate of a CD. These additions add to and change the intended sound outcomes of the playback devices and can, in part, be understood as a continuation of the "sound expansion" of the musical avant-garde, and as part of the opening up of music to all sound.

Once these practices are understood from an expanded understanding of sound and aurality itself, the interdisciplinary of these manipulated, cracked, and broken practices can be grasped. For example, key practitioner Nam June Paik has a history in composition but has become better known as an exponent of the moving image; and Christian Marclay is as well known as an experimental turntablist as he is a sound artist. As such the developing interdisciplinary area of sound studies provides a relevant context for the research of cracked media.

The Sonic Turn

In the years since the beginning of the twenty-first century there has been an increased theoretical interest in sound, in what Jim Drobnick has termed the "sonic turn."[15]

Michael Bull and Les Back begin their introduction to *The Auditory Culture Reader* by suggesting that we are no longer in a situation where vision dominates the manner in which we come to know our surrounds: "the experience of everyday life is increasingly mediated by a multitude of mechanically reproduced sounds . . . in parallel to this, cities are nosier than they ever were in the past."[16] Faced with this increased stimulation of our auditory senses we have been pushed into thinking in more detail about how we experience life through this previously underconsidered sense. How are we to relisten to the sound world around us, and how do we situate our bodies and our everyday life through these discourses?

Many recent publications in the area of sound culture move away from music and into sound. Veit Erlmann, writing in the anthology *Hearing Cultures: Essays on Sound, Listening, and Modernity*, argues for the need to focus on "extramusical sound," as there has already been a shift in focus in ethnomusicology and musicology from the reading and interpretation of scores to the materiality of musical communication.[17] Erlmann brings together scholars from different disciplines all of whom "share a recognition of the need for the cultural and historical contextualization of auditory perception."[18]

Christoph Cox and Daniel Warner point to an emerging audio culture, an "explosion of interest in auditory history and anthropology led by social scientists who have turned their attention to sound as a marker of temporal and cultural difference."[19] Drobnick in *Aural Cultures* argues that "Most pertinent to this publication is the embeddedness of sound in cultural, political, and physical contexts."[20]

The approaches taken in these recently published anthologies have been deliberately multidisciplinary. Those working in the field emphasize the importance of a diversity of approaches as many discipline areas begin to relisten to the sounds of their area of study. Erlmann explains that it is not possible for an anthology devoted exclusively to hearing to cover the wide range of "topics, methodologies, technologies, historical periods, and geographic areas awaiting further study."[21] Bull and Back make clear the breadth of interest in sound, with contributors to *The Auditory Cultures Reader* coming from diverse fields such as "sociology, cultural studies, media studies, anthropology, cultural history, philosophy, urban geology and musicology."[22] Drobnick states in his introduction to *Aural Cultures*, entitled "Listening Awry," that "As the parameters of sound increase, so too does the diversity of its discourses . . . it is quite conceivable now to assert that any discipline could profitably initiate one or more sub-disciplines devoted to sound within its domain."[23]

It is not only in academic circles that the "resurgence of the ear" has occurred. Within music and sound culture there has been a retuning and refocusing on listening closely to sound. Christoph Cox and Daniel Warner draw attention to a new audio culture that has emerged over the last fifty years that is "a culture of musicians, composers, sound artists, scholars, and listeners attentive to sonic substance, the act of listening, and the creative possibilities of sound recording, playback, and transmission."[24] This type of listening leads directly to a culture that desires knowledge of past sounds and past approaches to the creation of sound, and audiences who desire to hear new and unique sounds for themselves.

Cracked Media has as its point of difference from these previous publications an explicit interest in listening closely to the multifarious histories of creative sound—that is, sound that has been made deliberately for the purposes of creative and artistic practices. Rather than on the social or cultural history of experimental musics or sound, the focus is on situating current tendencies in experimental sound and the media arts. This approach is very different from the social and cultural approach taken in almost all these anthologies (the exception being *Audio Cultures*), which are not interested in the *art* of sound and spend very little time on the arts. Their approach is similar to that of cultural studies, in which popular and mainstream cultures are studied in an attempt to gain an understanding of the contemporary situation.

My approach is closer to the methodology utilized by Douglas Kahn throughout his publications on sound in the arts. Kahn has long been a champion of listening back through the arts in what he has called the "deaf century." He begins his book *Noise, Water, Meat* proclaiming:

> Sound saturates the arts of this century, and its importance becomes evident if we can hear past the presumption of mute visuality within art history, past the matter of music that excludes references to the world, past the voice that is already its own source of existence, past the phonetic task-mastering of writing, and past what we might see as hearing.[25]

By arguing that sound has been a constant and perhaps even disruptive element of the arts in the twentieth century, Kahn has written himself a discipline and made abundantly clear that we are no longer in a deaf period.

Cracked Media is interested in listening back through the history of the use of cracked media in sound practices with the goal of addressing the imbalance of theoretical and critical attention being paid to creative sound practices. In many ways this history starts where Kahn's history in *Noise, Water, Meat* ends, in the 1960s. It was at this point that musicians and artists truly started to crack and break the technologies of playback.

Cracked Playback Technologies

The focus of *Cracked Media* is the use of mediating devices for the generation of sound and performance practices. By mediating devices I mean technologies that have been designed and manufactured as playback equipment for recorded media. The uses of phonograph and compact disc technologies are the focus, as they display significant and sustained practice and development. The practitioner thus employs music's playback technologies to create new music, sound, and art through the modification, cracking, and breaking of the playback devices and their media software. Discussion will be limited to the use of mediating technologies for *sound* outcomes. This is important to note as the practice of cracked media itself can be witnessed in other areas of the arts including literature, visual media, moving image, and the diverse forms of digital media.

Sound recording technologies were used extensively throughout the twentieth century as both a compositional tool and a performance device. Cracked media forms a significant thread within this broader trend. Other media, such as analog tape, have in the course of the last century undergone cracked practices, but it is specifically the use of CD and phonographic technologies that will be addressed here. Analog tape is discussed in the section on Nam June Paik, but is otherwise not approached. This is due to the fact that the cracking and breaking of tape technologies was not a significant and sustained practice. Tape technologies were used as an important form of composition in the second half of the twentieth century, but for the most part this was achieved through the legitimate practices of editing and the extended practices of looping.

Although this study considers practices that use specific media technologies, it is not about technologies as such. It is not a study of technology driving creative practice—a form of technological determinism. Jonathan Sterne, in his book *The Audible Past: Cultural Origins of Sound Reproduction*, argues that technologies cannot be simply read through "impact narratives":

> In "impact" narratives, technologies are mysterious beings with obscure origins that come down from the sky to "impact" human relations. Such narratives cast technologies themselves as primary agents of historical change:

> technological deification is the religion behind claims like "the telephone changed the way we do business," "the phonograph changed the way we listen to music."[26]

Technology does not appear in a finished state ready to be passively consumed; instead it is in a constant state of development in the market and the domestic sphere, as well as in historical and theoretical discourses. To put it another way, Timothy Taylor, in his book *Strange Sounds: Music, Technology, and Culture*, argues that there are problems with both the determinist model and its opposite, a purely voluntary model:

> The slipperiness of the term *technology* can lead to its reification, lifting it out of the social, cultural, and historical webs in which it is produced and used. One would have to have been living outside our current moment not to have heard phrases such as "technology changes the way we see . . ." as though it were technology and not ourselves making and using various technologies in a complex series of interlocking webs.[27]

Statements such as "technology changes the way we see" miss the social and cultural significance of technologies and the way the social and the cultural shape the manner in which we use the technologies themselves.

The manipulation of mediating technologies in cracked media calls into question theories of the effects that recording technologies have had on music. Mark Katz writes in his book *Capturing Sound: How Technology Has Changed Music* of "phonographic effects," which are responses from musicians, composers, and audiences to the phonograph and recording technology in general.[28] Katz is not arguing, however, that the phonograph forces or makes one respond in a certain way or changes the way we respond to music. He states:

> I myself write of recording's *influence* on human activity and of phonograph *effects*, both of which impute casual powers to technology. Although we often respond to technology within a context of limited options not of our own making, we must remember that in the end, recording's influence manifests itself in *human* actions . . . the influence

> that I describe does not flow in one direction only, from technology to user.[29]

The practitioners of cracked media further support a nondeterministic understanding of the effects of technology and illustrate the two-way flow between the user and technology. Mediating technologies, such as the home stereo, have a singular intended purpose: to play back music. The reason we have a CD player in our living room is to listen to recorded music; it is there to mediate between the data stored on the software (the actual CD) and the sound played through the speakers themselves.[30] How and why these technologies have been used to actually create new transformative sound, and how it is that these devices can be heard to create and extend the sound world, are key questions addressed here.

At the center of the discussion is the concept that computer mediation of digital data erases the specificity of media: data itself is abstract and can easily be transformed from one media form to another. For example, data intended to be visual can be played through a sound engine. In the mid-1980s this consequence of digitization was foreseen by media theorist Frederick Kittler, who proclaimed:

> The general digitization of channels and information erases the differences among individual media. Sound and image, voice and text are reduced to surface effects, known to consumers as interface. . . . Inside the computers themselves everything becomes a number: quantity without image, sound, or voice. And once optical fiber networks turn formally distinct data flows into a standardized series of digitized numbers, any medium can be translated into any other. With numbers everything goes.[31]

In a computer-mediated environment, data has no essence or true intent. The final projection, visual or aural, is not contingent on the actual nature of the original data. Media and mediation are never transparent. Mediation actively transforms data from one form to another and is never passive. Moreover, the active transformation of media is not solely linked to digital media. All mediation transforms information actively, and cracked media practices make this clear as the site of mediation. The record and its playback technologies, for example, can be used to creatively transform their intended outcomes. Cracked medial approaches to production is an active creative strategy that explicitly draws on a critique of media and mediation itself.

These practices stand in direct opposition to the rhetoric that surrounds the home entertainment industry and its drive for high-definition vision and sound, media that strive to create a transparent experience that transcends mediation. This rhetoric literally made it possible for us to listen to recorded music. Jonathan Sterne discusses the rhetoric of the early history of reproduction and fidelity during the technology's early development:

> Sound fidelity was, ultimately, about faith and investment in these configurations of practices, people, and technologies. It posited the technology to reproduce sound as a vanishing mediator. . . . They had to be convinced of the general equivalence of the live and the reproduced. Even when the *sounds* of sound-reproduction technologies were explicitly discussed, it was with an eye toward finding new ways for the medium to erase itself.[32]

Listeners were asked to ignore certain aspects of the technology itself, such as the graininess of the wax and its dull frequency range, so that they would be able to hear the "real" sound of the recording. These leaps of faith led to a belief in the system and in very real terms allowed it to become a usable technology given the very great audio limitations of the early versions of the device.

The function of the sound itself was contingent on the belief that mediation would, at an end point, disappear, allowing a truly transparent perfect copy of the original live event. Sterne discusses the "mediation thesis" as positing a moment of "unmediated sonic reality prior to sound's technological mediation."[33] Subsequently an ideal form of mediation is imagined as being fully transparent, "where the medium produces a perfect symmetry between copy and original and, thereby, erases itself."[34] This goal has not, of course, been reached, and in fact through the practice of cracked media we see that mediation itself has become the object of sound creation, composition, and performance.

This rhetoric of transparency, as well as the permanence of the technology and the actual recordings, is directly critiqued by the practice of cracked media. The one-way flow of data to the passive recipient is displaced in a practice that uses this media to create original, transformative sounds and compositions. The fact that these finished or final recorded products and playback technologies have been taken and folded back into themselves, used as machines for transformation into new practices and works, is of key importance.

These practices usually occur in two forms: the live performance—a one-off event that cannot be repeated; and the recording—a "rerecording" of the cracked and broken mediating tools. These twists and folds will be discussed later on in greater historical and analytical depth.

Manipulated, Cracked, and Broken

The approaches that have been taken by the practitioners of cracked media are heterogeneous in nature. At the extreme end is a destructive approach in which the mediating device or recording is broken, perhaps beyond repair. Toward the center is the less finally destructive crack, where impermanent or nonfatal damage is caused to the player or the recording media. The least extreme approach uses the simple manipulation of playback technology—sometimes an extremely subtle intervention.

Manipulation is the extension of the technology beyond its originally intended use. Many of the cracks and breaks are not literal. That is, they often take the form of a temporary disruption to the steady flow of recorded data, rather than a permanent crack or actual break. This is often the case in glitch; many glitches were created by temporary problems with the CPU or with the applications being used, and on restarting the computer the "problem" no longer occurs. The technology is bent, almost to its breaking point. We might think of a young student bending a plastic ruler, almost to the point of snapping. The student wants to see how far he can push the ruler but does not want to actually break it. Manipulation is initially very safe, with no long-term damage caused to the tool. Early turntable manipulation is an example of this. Composers took the turntable and used it beyond its intended purpose by letting the turning mechanism of the platter wind down, by playing the record at incorrect speeds and by playing the record backwards. This type of practice typically happens very soon after the technology is released onto the market, as its users work out what can be done with it and how far it can be driven before breakdown. One does not want to break his or her newly purchased phonograph or record, but the artist is interested in seeing how far it can be pushed and pulled—how will it sound if I try this or that?

The manipulation of mediating technologies quickly develops into cracked technology. In the previous analogy this means bending the ruler just a little too far and forming a weak spot or small fracture. The cracked object is certainly damaged, but it is still usable. The cracked jug will still hold water, but it will always be in less than mint condition.

Perhaps the crack is a little more severe and this means that the jug will no longer hold water, but the jug is still kept as an object of nostalgia, carefully placed on the mantelpiece for display. The point is that we still hold onto the object; it is not simply discarded. Similarly, cracked media may be useless for its intended purpose (that is, the seamless and transparent playback of recorded media), but it is still useful for the creation of new sound and performance techniques. The sounds produced through cracked technology are filled with noise, as unintended and extramusical sounds are pulled from the technology as it is pushed to the edge of breaking.

The temporary crack is closest in intent to manipulation and is closely related to the impermanent preparation of musical instruments. These cracks resemble modification of the technology but with the intention of generating breakdown or noise rather than the extended or improved use that is the usual rationale for modification. The media will perform other than intended while it is cracked, but it can be fully restored to its original form after the event. As mentioned earlier, this could include applying tape or impermanent marker to the substrate of a CD; the technique does not harm the system in any way, and in the end the CD can simply be cleaned and returned in its original condition.

Closer to the break is the permanent crack, an example of which is the scratched record. Records are delicate objects, which can easily be scratched; once a record has been scratched, the tick or pop caused by the needle hitting the mark on the vinyl surface can never be removed, and the data flow is permanently damaged by the mark. Depending on the extent of the damage it is likely that we will keep the record, even though the pop might cause annoyance on playing. The record is still playable; it still does the job it was intended to do, although without the possibility of transparency. The deliberate scratching of a record, or the deliberate use of a record that has been scratched or scuffed, is a more extreme example of cracked media.

Damage done to recorded media or playback devices, in the form of the crack, takes two distinct forms: accidental damage and deliberate damage. The former is caused by general use and might be manifest in, for example, the found record object. Perhaps the vinyl album was discarded and picked up by the artist, or the record was already in use in the artist's collection and is now heard as being ripe for the creation of new cracked sounds. The latter approach is that of deliberate damage, where, for example, a sharp object is used to deliberately damage the surface of the record, or perhaps the record is deliberately scuffed by placing it on the floor of a gallery. Here we have two distinct intentions, but both can lead to further and more extreme damage.

The deliberately damaged recorded object often leads to more extreme damage and finally to the complete destruction of the object. Once something has been devalued by a permanent crack it is more easily destroyed. The slightly damaged object is of less value than the pristine, newly acquired technology, and this object of reduced value quickly begins to slide toward total destruction. This path, often accelerated by the obsolescence of the technology, causes the media to become expendable and therefore ripe for destructive experimentation.

The most extreme practices of damage and destruction are referred to here as *broken media*. Musicians have smashed and broken their way through numerous turntables and media including vinyl records and CDs. This destruction is often of a physical, performative nature, with records snapped during performance, cymbals crashed on top of tone arms and cartridge heads, needles smashed into the platters.[35] The theatrics of such tactics are clear, and the audience is often shocked by such extreme destructive acts. The visually performative aspects of this destructive tactic connect to an often extreme audio outcome, which might take the form of a dense and complex resonant feedback loop, played at extreme volume, or the crack and snap of shattering vinyl.

The use of extreme destruction in the work tends to lead to a high level of chance and chaos, not least because of the unknown nature of the break and the audio created out of it. This destruction can lead to physically dangerous work such as the practices of Lucas Abela as described earlier. Abela has been admitted to the hospital with a severed radial nerve after a performance swung to the physical extreme.

Media Archaeology and Cracked Media

This book is an examination of previous ways of using technologies that may, on the surface, seem very different from current technologies (many of the technologies looked at are predigital). It reads a prehistory through these practices and through the retooling of earlier technologies for outcomes that foreshadow current practices. This bears similarities to the emerging field of media archaeology, an increasingly prevalent approach to the study of the contemporary media arts.[36] *Cracked Media* develops an understanding of contemporary media and the manner in which it is used, through its predicates and precursors. By approaching current trends in the use of media in such a way we can reconstruct the various developments in media history that led to the current situation, and by doing this we bypass the problem of a technological determinism that sees technology as driving practice rather than practice and use driving technology. It also avoids the problematic celebration of new technology in which the novelty of the technology itself is the focus. Here we understand that all technologies have a web of histories associated with them.

Currently we are experiencing the rapid growth of a new type of home entertainment via the almost total convergence of all media through digital mediation and the personal computer. The homogenization of applications and approaches across media has changed the way we approach music, film, and photography. We are still consumers but have also become producers. Apple Macintosh has accelerated and popularized this shift by integrating and simplifying software applications used for production of media, providing easy-to-use, transparent, and accessible user-interfaced tools in the package "iLife," which includes the applications iTunes, iDVD, iPhoto, and iMovie.[37]

In this era of domestic digital media, the lines between creator, producer, programmer, and consumer are blurred, and new modes of creation are no longer the domain of the genius artist. Anyone with home computer equipment is able to create his or her own environment surrounded by media of his or her choosing and creation. The computer is at the hub of this "meta-media" environment, and it acts as the central device for the convergence of media. As Lev Manovich asserts:

> Today it is the meta-media paradigm—rather than other seemingly original computer techniques or even computer programming itself—that is at the center of computer culture. Outside of the professional and scientific domains, consumers think of computers as machines for downloading, storing, transmitting and editing media. Software such as media players and editors, CD and DVD burners, ports to connect digital still and video cameras or MP3 players are what consumer computing is about today.[38]

This represents a marked shift from the notion of the passive consumer of home entertainment, of consumers slouching on the couch in front of their television, staring blankly at the light-emitting box in the corner. Now we actively hunt out the content of our entertainment by downloading it from the Internet or from our digital cameras, editing it into a form that we feel best suits our individual purpose. Whether creating short movies of our holidays for the family or editing a library of literally thousands of music tracks for our iPods, the way in which we interact with our playback equipment through the use of computers is very different from the almost reverential approach to, for example, the handling and playback of a vinyl record.

The convergence and subsequent user development of computer-based applications did not develop out of nowhere: again, this is not a case of technology changing the way we live our lives, but yet another case of users directing the development of new technologies to suit their purposes. Apple can be seen in part to be filling a niche, and, as I will demonstrate, the practices of cracked media represent an early developmental phase of this current moment. These artists took the passive and transparent tools of playback and manipulated, cracked, and broke them into new forms for creating original compositions, unique performance tools, and new instruments. What all the musicians and artists in this research have in common is that they took their home entertainment equipment, their turntables and their CD players or the vinyl records and CDs, and used them creatively beyond their initial intended purpose.

This book is divided into four chapters. The first chapter, "Recording and Noise: Approaches to Cracked Media," discusses the nature of recording technology and the music industry in relation to the crack and break. It will be seen that the flow of production and consumption is transformed as practitioners produce new and unique sound practices out of what were intended to be sealed, finished products. This chapter will also pay close attention to noise and discuss its various manifestations in relation to the practices of the crack and break. It will be seen that neither theories of recording nor those of noise provide a fully realized framework in which cracked media can be understood, and therefore other approaches need to be sought.

The subsequent two chapters, "Broken Music: The Manipulated, Modified, and Destroyed Phonograph" and "Damaged Sound: Glitching and Skipping Compact Discs in the Audio of Yasunao Tone,

Nicolas Collins, Oval, and Disc," will delve deeply into the history of the crack and break, looking at exactly what was being produced from early in the twentieth century through to more recent experiments in the area. Two distinct approaches to cracked media will be discussed at length: those that use the phonograph and those that use the CD.

Finally, in the chapter entitled "Tactics, Shadows, and New Media" three possible frameworks are proposed for the further understanding of cracked media practices, in relation to contemporary uses of new media. The first part of this chapter will discuss cracked media in terms of Michel de Certeau's tactics of everyday life. Cracked media utilize everyday technologies and redirect them toward transformative practices in a manner similar to that outlined by de Certeau. Second, a discussion of the "shadow effect" of technology will look to the practice as a shift in the nature of foreground and background. Here the sounds that are usually hidden underneath the content of any recording are brought to the fore. Finally, the book will show how cracked media can be understood as precursors to the now commonplace practices of computer-driven "meta-media." It will be demonstrated that cracked media practices open up previously closed "old media" forms, transforming them in a way that closely resembles current open and modular new media forms.

1 recording and noise

approaches to cracked media

Introduction

The technologies of music reproduction are designed to be the end point of the recording industry, but in the hands of practitioners of cracked media they become the basis for the generation of original sounds and performance. Given their basis in recording technologies, it would be logical to consider critiques of the recording industry as possible starting points for the further discussion of cracked media. Initially this discussion will consider the work of Theodor Adorno and Jacques Attali, whose negative critiques of the recording industry are frequently cited in contemporary discussions of these issues.

The discussion will raise questions such as: Do the practitioners of the crack and break react against the recording industry? Is it possible to extend the commodities of recording beyond capital? The practice of cracked media problematizes the negative critiques of recording proposed by Adorno and Attali, as the very act of recording becomes an originating creative act.

Next, the discussion turns to discourses of noise. Again, it seems sensible to look directly at theories of noise, which may then be applied to the practices analyzed in this discussion. The sounds produced by the musicians and artists considered here are, after all, based in noise. These practices are filled with noise—quiet and loud; gentle and destructive; moderate and intense. If noise is at the core of the practices of broken and cracked media, then it is necessary to locate what noise means in terms of these practices and to consider whether existing theoretical approaches to noise can be utilized for this purpose.

Noise theory, however, is itself chaotic and filled with contradictions, and as such provides an unclear path. The practice of using cracked and broken mediating devices complicate some of the reductive views on noise and also offers a major theoretical critique of noise itself.

1 The Critique of Recording Technology

> Talking machines and phonograph records seem to have suffered the same historical fate as that which once befell photographs: the transition from artisanal to industrial production transforms not only the technology of distribution but also that which is distributed.
>
> —Theodor Adorno, "The Curves of the Needle"[1]

In the quote above, Adorno commences a discussion on the development of recording and playback technology, and more specifically the changes brought about by new materials that "wear out faster than old ones."[2] For Adorno, new methods of distribution (musical recordings) actually change what is being distributed (music itself). He understands new musical technologies to actually devalue the live experience of music, replacing it with inattentive private listening. In a similar way, Jacques Attali in *Noise: The Political Economy of Music* positions new recording technologies as an example of society's shift toward capitalism and by association mass production, turning music into a mere simulacrum in the process.[3] For Attali, studio simulations of live performance (the studio recording) have caused actual live performance to disappear into simulation. The only hope for music, then, lies in its future "Composition" phase. Here, Attali argues, it will be possible to have "a revived radicalism of constructivist noise or athematic 'informal music,' all accompanied by progressive social claims."[4]

For both Adorno and Attali the shift in music caused by the technologies of recording and playback is a negative one, one that devalues the live, "real" experience of music. Although these theorists are writing in very different periods, with different agendas, their basic assumptions about the nature and effect of recording technologies are quite similar. Adorno and Attali view recording technology as a symptom of capitalism, transforming the ephemeral nature of music into something solid, something that can be bought and sold. This is understood as a negative effect of capitalism, alienating the producer and consumer and also turning recordings themselves into items of fetish.

Adorno and Attali are certainly not the only critics to discuss this alienating effect of recording technology. Evan Eisenberg, for example, in *The Recording Angel* discusses the "music industry" in a chapter tellingly entitled "Music Becomes a Thing." He explains:

> When I buy a record the musician is eclipsed by the disk. And I am eclipsed by my money . . . when a ten-dollar bill leaves my right hand and a bagged record enters my left, it is the climax. The shudder and ring of the register is the true music; later I will play the record, but that will be redundant. My money has already heard it.[5]

Eisenberg discusses the idea of fetishism in relation to recorded music, pointing out that the product of the musician's labor is removed from the musician as they themselves become consumers of recordings, as opposed to delivering live concerts and performances, and as such the product of their labor is beyond their reach. This alienation of labor, in the Marxist sense, works both ways: the musicians may never see the consumer, and the consumer may never see the work that led to the recording. For Eisenberg, this leads to the situation where the insatiable consumer of recordings desires the purchase of the record more than the listening experience that happens after the exchange: "money wants to be spent, and if I fancy myself a music lover it seems natural to spend it on a record. I could spend it on a concert, I suppose—but a record is tangible, like money."[6]

Adorno and Attali and the Negative Effects of the Recording Industry

Recording has arguably had a democratizing effect on music. Put simply, the recording of music has provided access to different kinds of music, to many more people, than otherwise would have been the case. The strongest example of this is radio. For the relatively cheap price of a radio a whole family can access the concert hall as well as music from around the world. Because of the reproduction of music, all strata of society have access to all types of music, including musics that had previously been out of reach of most owing to the expense of witnessing, for example, the opera, or the difficulty of experiencing musics from remote locations.

The most commonly cited example of the argument for reproduction having a democratizing effect on cultural forms comes from Walter Benjamin's "The Work of Art in the Age of Mechanical Reproduction." Benjamin shows why film (explicitly) and recording (in passing) have the effect of opening culture up to the masses and are the art of the masses. The argument can be found initially in this statement:

> technical reproduction can put the copy of the original into situations which would be out of reach for the original itself. Above all, it enables the original to meet the beholder halfway, be it in the form of a photograph or a phonograph record. The cathedral leaves its locale to be received in the studio of a lover of art; the choral production, performed in an auditorium or in the open air, resounds in the drawing room.[7]

The consequence of these spatial shifts is that elements of culture that were previously unattainable can be accessed by the masses.

The choral production, once available only to those who could afford a ticket, can now be heard in the drawing room, most probably initially through listening to the radio. It is the very reproduction of the artwork that "emancipates the work of art from its parasitical dependence on ritual."[8]

Adorno, however, does not perceive this as democracy or as a positive development. In his essay "The Radio Symphony: An Experiment in Theory," he makes his position on the negative effects of radio, and by association the phonograph, very clear:

> We are primarily concerned with pointing out the fact that serious music as communicated over the ether may indeed offer optimum conditions for retrogressive tendencies in listening, for the avalanche of fetishism which is overtaking music and burying it under the moraine of entertainment.[9]

Adorno, in "The Radio Symphony," compares the live performance of a symphony to that of a recording of the piece broadcast via radio transmission, arguing that: the intrinsic nature of the live performance is bound up in variation, an aspect that cannot be reproduced via the medium of recording; the "absolute dynamics" of the concert hall are not adequately reproduced via recording; the volume of the work in the concert hall and the ability to "enter" the work is not reproduced in the privacy of the home; and the subtleties of the symphony are lost to reproduction. To Adorno's ears, the works are transformed into poor chamber music.[10] He concludes from these arguments that the radio in fact has no true democratizing effect; rather what is actually presented is not the work but a distorted shadow of the original.

Adorno's critique was strongly influenced by the state of recording and playback technologies at the time. Fidelity was poor and long-play records were yet to be released on the market, and as such the actual experience of listening to records, or the radio, could not compare to that of attending a concert. Nevertheless, Adorno's critique is still valid in the contemporary context where fidelity is much better, as his argument is about the democratizing effect of recording technology. Even perfectly transparent playback cannot reproduce qualities of live performance such as variation, chance, and environment; hence the masses are still getting a poor copy of the real thing.

Michael Chanan extends this argument in *Repeated Takes: A Short History of Recording and Its Effects on Music*, observing that music can now be heard anywhere a sound system is set up and that with this shift of reproduction comes the dispersal of community.

Instead of small like-minded or socially similar communities who were entertained as a group in their own localized vernacular, audiences can now listen to music of different cultures, countries and communities at the "touch of a button," and "the result is that musical experience has been radically altered."[11] Chanan reads this alteration of music through reproduction as leading to the alienation of the musician and the listener from music, as labor and consumption are alienated from the commodity.

Adorno's critique also directly addresses the phonograph in the two essays "The Curves of the Needle" (1927) and "The Form of the Phonograph Record" (1934). As quoted at the beginning of this section, Adorno perceives the phonograph as suffering the same fate as photography, whereby the art (here music) suffers through its transition to industry. In "The Form of the Phonograph Record" he dismisses any artistic potential of the phonographic disc as its very structure, its form, hinders artistic potential. It is clear from these words that he did not know of the then-contemporary experiments that used and misused the record and the phonograph to produce new phonograph-specific music. For him the record was "the first means of musical presentation that can be possessed as a thing. . . . records are possessed like photographs."[12] Music had become a commodity, an object to be owned, collected, archived, and stored away. This is a major shift in the history of music, a shift that dramatically changes the form of music as it is molded to fit its new place in capitalist society. The record object is the mediating device between production and consumption. This mediation, however, does not simply flow in one direction. Instead the act of consumption changes the actual production and performance of music, as music is now created for the consumption of the record object. If, as Adorno argues, recordings are shadows of the real act of music, its performance, then the creation of music for the recording industry is the deliberate creation of a debased musical object. This is an inevitable negative effect of recording, and the impact on music grows as the recording industry becomes more significant within society.

Jacques Attali takes this argument to a further extreme, bringing both Guy Debord's "spectacle" and Jean Baudrillard's "simulacrum" into the critique of the production of recorded music. In *Noise: The Political Economy of Music*, Attali uses Marxist historical determinism to discuss the development of music. Music is, for Attali, a prophetic force, as it foreshadows developments in capitalism and history before other material effects are perceived.[13] He demonstrates how this has occurred in various phases of history, the key phase for this argument being the "Repeating" phase. Through the recording process and the distribution of recordings,

> the relation between music and money starts to be flaunted . . . more than ever music becomes a monologue. It becomes a material object of exchange and profit, without having to go through the long and complex detour of the score and performance anymore. . . . Once again, music shows the way: undoubtedly the first system of sign production, it ceases to be a mirror, an enactment, a direct link, the memory of past sacrificial violence, becoming a solitary listening, the stock-piling of sociality.[14]

For Attali the development of a resource that can be stockpiled and music's new relation to consumption and capital lead directly to the situation where the process of commodity consumption reduces music to a simulacrum of its original. Thus:

> *the growth of exchange is accompanied by the almost total disappearance of the initial usage of the exchange.*

> Reproduction, in a certain sense, is the death of the original, the triumph of the copy, and the forgetting of the represented foundation: in mass production the mould has almost no importance or value in itself.[15]

What was originally represented in the recording—the performance—is lost in layers of simulation. As this loss occurs, we can argue, live performance completely vanishes; the simulacrum has no grounding, no actuality in reality. Instead the studio becomes a simulation itself of performance, and in turn the recording a simulation of that simulation.

Attali, similarly to Adorno, does not see reproduction as the emancipation of art; instead he sees it as a force utilized by capitalism, shifting performance from the public to the private sphere before turning the performance itself into a form driven by manufactured simulation. The technology itself, however, can be pulled to the front of the creative experience, as argued by Greg Hainge, who develops Attali's theories in a paper entitled "Come on Feel the Noise: Technology and Its Dysfunctions in the Music of Sensation." He describes his music collection, one that is filled with noise recordings. This collection includes: Merzbow's fifty-CD boxed set *Merzbox*; Jazzkammer's electronic circuit noise; Reynol's *Blank Tapes*, which, as the title suggests, is a recording of blank tapes; Francisco Lopez's *Paris Hiss*, which is similarly recorded; and finally the "ambient" glitch of Oval and Pole. Hainge asks:

> What happens . . . when the primary content of the sound processed by a high-fidelity system is composed precisely of those sounds that the system is designed to eliminate? . . . What happens at the level of reception, however, is very different to the normal processes of musical receptivity, and this is due . . . to this "meta-noise," to the elevation of mechanic acoustic by-products to the level of primary content.[16]

Developing Attali's repetition phase of music, Hainge argues that repetition is only made possible through technology that allows for an infinite number of copies to be made, and these copies, though resembling the original, in fact lead to simulacra. These technologies are turned toward the development of the perfect copy of a music that no longer has an original. Noise musicians, however, highlight those elements of music production that the production process attempts to silence, and as such they foreground the technological system itself. He argues, "In creating music from the sounds of the dysfunctions of that technology, they focus our attention on the fallibility of the systems that we have constructed and in which we believe, breaking their transcendent possibilities and ensuring the creation instead of a direct affective aesthetic."[17] These shifts occur within the technology and as such form a tactic by which musicians can utilize the technologies of recording toward an end that is not aimed at repetition and simulation, but one that foregrounds the technology itself. The use of noise within recording is a comment by the musicians as to the "medium's role as mediator within an era of repetition . . . we might say that Noise is the sound of technology's spasms as it attempts to escape itself."[18]

Cracked Media and Critiques of Recording Technology

The use of cracked media in the creation of sound and music problematizes Adorno and Attali's critiques of recording technology. It calls for a reevaluation of the potential for original and creative output from the end point of the recording industry, the point of playback. First, the flow of production and consumption is disturbed by the productive musical outcomes generated by cracking and breaking media, and second, the fetishistic character of musical consumption is questioned by the abuse of the reified products of the music industry.

The practice of cracking and breaking playback media folds the flow of production and consumption back on itself. The imagined transparent and passive mediating devices of storage and playback are transformed into generative technologies by practitioners of the crack and break, breaking the linear flow of production and consumption.

The manipulators of playback technology are not simply consuming products, in this case prerecorded music. Instead they are actively using the end products to create original sound and compositions. In turn, the sound and performances created through this retooling of playback data are themselves recorded, often (in recent practice) to CD-R or MP3. CD-Rs and MP3s are readily traded; their value as a commodity is not monetary, but rather is found in the development of status for the artist, often only within a very small group of friends or like-minded artists. These techniques strip the products of recording of their alienating qualities, bringing the labor and the product of labor back together and clearly challenging the simple one-way production–consumption model.

With the breakdown of alienation comes the breakdown of the fetishization of recorded music. Musicians at the center of the use of recording technologies for productive creation use the crack and break to transform the recording. This is sometimes carried out through destruction of the media of music playback (the vinyl record or CD). Through the destructive act, artists directly question the value of the media itself and the way we have been taught to carefully handle it. This act contravenes any reverence held toward the fetishized object and thereby challenges Adorno's view of the fetishization of music and the value of consumption itself, as it does Attali's idea of the spectacle of music and its very consumption.

2 What? I Can't Hear You Over That NOISE!

Noise is filled with "complexity and unpredictability" and is a particularly desirable area for the experimental musician interested in new sounds and possibilities for performance. The sounds of noise are vast and varied; it is just as likely to produce deafening rupture, chaotic static, and screaming feedback as it is to generate extremely quiet sounds, subtle clicks and snaps, and vinyl crackle and pop. Although initially noise appears to be the perfect direction from which to approach the practice of cracked media, it soon becomes clear that many theories of noise do not adequately address the subject. Perhaps the practice itself calls into question conventional understandings of noise.

Noise fills the audio output of cracked media: cracked lines, lost data, static and hiss, broken signals, chaotic production, earth hum, piercing tones, and digital glitch. All these sounds are made up of what we call noise, and many of the approaches taken toward the crack and break seem to fit into the numerous definitions of noise. Noise, however, eludes simple definition. Noise has been theorized in discrepant disciplines such as information theory, acoustics, musicology, and "everyday meanings,"[19] yet none seems adequate to fully account for noise in music.

There is also a need to distinguish between aural noise (the sound of noise) and theoretical readings of noise (conceptual or abstract noise). Noise theory of recent years has focused heavily on Japanese noise music, a genre of music that comprises extreme sonic practices. This focus causes numerous complications in the understandings of noise, including the very possibility of the seemingly self-contradictory genre of noise music. Noise theory looks to noise as a disruptive and excessive area of sound practice and finds within it a joyful transgression. Following a discussion of theoretical readings of noise, this section will relate these ideas to the practices of cracked media. However, the practices of cracked media present more pitfalls for these theories; in the final reading the practice exists largely beyond this area of theory.

Noise Theory in Practice

Noise theory is full of excess, making bold, semipoetic statements about noise's ability to displace calm and prolong "disquietude by opening up the divide between crisis and restoration, certainty and uncertainty."[20] Noise is often heard as excessive and transgressive, as being loud and disruptive. The theory is also often caught up in this excess. David Cunningham in "Goodbye 20th Century: Noise, Modernism, Aesthetics" alerts us to this connection and its accompanying reaction:

> *noise* as a specific figure of *excess*—with its accompanying metaphorics of the "ear-splitting," the "over-powering," or even the "unlistenable"—has often seemed particularly liable to provoke a concomitant *rhetorical* excess on the part of its would-be theorists.[21]

Noise is widely considered to be situated within excess, as a transgressive act that exceeds managed data. Noise is "out of control," and as such its theorists are pulled into its chaos with ringing ears. The excessive and seemingly transgressive nature of music that utilizes noise as a key component of its content is exemplified in the ultraextreme subgenre of noise music, Japanese noise. Discussions of Merzbow (Akita Masami), the leading figure in the subgenre, are full of these trends.[22]

Japanese noise music is a subgenre of noise music and is the most excessive and extreme version of the genre to date. Centered in Osaka and Tokyo, it can be divided further into power electronics (musicians include Merzbow, Aube, Masonna, and KK NULL), and psychedelic and free noise (musicians include Haino Keiji, Hiroshige Jojo, and Incapacitants), with numerous permutations occurring in between. Power electronics has received the most attention from the academic community. The subgenre itself was most active from the early 1980s through to the mid-1990s, after which it dissipated, with only a few remaining exponents continuing to produce music in this style.

The music created by the two most frequently cited noise musicians, Merzbow and Masonna (Yamazaki Maso), is produced to be heard at massive volume in a barrage of noise. The music utilizes analog feedback as a central device and produces sounds grouped around low-frequency throb and screeching high end. Merzbow and Masonna could not be more different in performance, however: in his contemporary practice, the quiet and introverted Merzbow sits in front of two laptops, staring blankly through his sunglasses at their screens. Masonna, on the other hand, makes a massive display of condensed energetic ruptures in a violent and chaotic performance, before either breaking something or hurting himself, bringing the brief performance to an end.

For Nick Smith, noise music is a genre that attempts to be dissonant with contemporary consumer culture.[23] This dissonance is sought via a number of tactics, including: exceedingly brief performances (Masonna performs sets as short as forty seconds); the release of large numbers of individual recordings (Merzbow's fifty-CD boxed set is the perfect example of this); and most obviously "unmistakable" music. By seemingly driving all commodity value out of the production of this music, Japanese noise artists go against the dominant capitalist trends in contemporary society. They also resist the analyses of Attali by subverting the connections between performance and recording, as well as the market for stockpiled recorded product. Smith argues that the tactic does not work, however, as "Once noise is no longer inscrutable . . . it is assimilated into popular culture and becomes commercial novelty."[24] Smith posits that "Masonna's cultural reception thereby demonstrates the mechanism by which modernity absorbs artistic attempts to critique it, and noise is ultimately understood as a desperate but spectacular failure."[25] It is partially the result of its assimilation in the mid-1990s that Japanese noise music is no longer a developing form.

Given the extreme nature of Japanese noise music, it is clear that this is not the most useful area to look to in discussing cracked media. This is, for the most part, the result of the excess of Japanese noise and the perceived desire for transgression. Cracked media practices are often neither excessive nor transgressive. In terms of volume, for example, they can be quiet or even barely audible. The subtle manipulation or the pop of a scratched record do not compare with the extreme nature of Japanese noise. Practitioners of cracked media are, however, part of the historical trajectory for noise that began with Luigi Russolo, and as such the use of noise in experimental music is a possible approach to further elucidate these practices.

Historical Noise: Russolo to Cage

As illustrated above, Japanese noise music has been widely regarded as the most extreme articulation of noise in music. It can be slotted into a trajectory of noise in music that began with the futurist call for the use of modern technologies in arts practices. There are numerous pre–World War II composers who have written influentially about noise in music, often as a call for music to allow noise to flood in. Although this line of history has been discussed by many authors in recent years, it is nonetheless useful to present a brief outline of this history here, as it provides context for future work in the area of cracked media.

The most obvious starting point for noise in music is Luigi Russolo's manifesto *The Art of Noises* written in 1913.[26] In it Russolo, in an overtly futurist fashion, hails the modern noise of the industrialized urban environment, calling for these noises to enter music:

> For many years Beethoven and Wagner shook our nerves and hearts. Now we are satiated and we find far more enjoyment in the combination of the noises of trams, backfiring motors, carriages and bawling crowds than in rehearsing, for example, the "Eroica" or the "Pastoral."[27]

He concludes the letter by calling for futurist musicians to "enlarge and enrich the field of sounds."[28] This was to be done by substituting the limited sounds produced by the orchestra with the infinite sounds produced by noise. Russolo's idea of noise is initially heard in the machines of modern industrialization.

Russolo's manifesto and his book published in 1916 were highly influential on a number of composers of the time, including Claude Debussy, Igor Stravinsky, Darius Milhaud, Arthur Honegger, Edgard Varèse, and Henry Cowell.[29] He is also now recognized as a key precursor to much of the "noisy" experimental music produced in the second half of the twentieth century.

Historical noise, here defined as the set of sound practices initiated by Russolo, encroached on Western art music and culminated in the noisy, Cage-influenced Fluxus movement. We can certainly hear noise in the commotion of Dadaist performances, and we hear the raucous joy of noise in the music of composer Henry Cowell. Cowell writes of his enjoyment of percussion instruments and the punctuation they produce through cymbal crashes and bass-drum rolls, and he calls for the further employment of noise in music as heard in the compositions of Edgard Varèse.[30]

It is John Cage and his influential paper from 1937, "The Future of Music: Credo," that truly signifies a filling out of the sound spectrum and of possible sounds available for use in music. He asserts:

> Where ever we are, what we hear is mostly noise. When we ignore it, it disturbs us. When we listen to it we find it fascinating. The sound of a truck at fifty miles per hour. Static between the stations. Rain. We want to capture and control these sounds, to use them not as sound effects but as musical instruments.[31]

In the paper Cage predicts a future music that will utilize all sound for the purpose of music. This future will use electronic means to create a music that will be much richer for its extended materials.

Cage's music is filled with noise, including the noise of cracked media. His formative influence on the use of cracked media is no more clear than in regard to the subsequent Fluxus movement. It was the Fluxus movement of the 1960s that took on and expanded Cage's exploded view of the field of music, and as a movement it was the first to truly crack and break media using the noisy process of sound creation.

Given the importance of noise in the developments of experimental music in the twentieth century, it is necessary to integrate these historical approaches with a theoretical understanding of the nature of noise. It will be shown that noise signifies an abundant source of material in that it is full of possibilities (perhaps all possibilities) and aids in furthering the understanding of the potential of sound practices. Noise is complicated, however, and a number of possible understandings of this area may be further utilized in determining the intentions of producers who use cracked media in their sound work.

What Is Noise?

That noise is a difficult concept is an understatement. Conceptions, much like noise itself, are often filled with misconceptions and discrepancies. There are at least four distinct categories for defining noise: acoustic noise, noise in information theory, subjective noise, and material noise. Noise, however, is not at all easy to tame, as it traverses the boundaries between these four categories in chaotic and insensible ways, remaining somehow neither/nor.

What are the properties of noise as a purely physical phenomenon? What is the difference between noisy and non-noisy sound? Herman Helmholtz, in his seminal text *On the Sensations of Tone*, describes the difference between noises and musical tones: "The soughing, howling, and whistling of the wind, the splashing of water, the rolling rumbling of carriages, are examples of the first kind, and the tones of all musical instruments of the second."[32] Put simply, irregular vibrations of the air constitute noise, whereas regular vibrations produce tones. Helmholtz continues, "The sensation of musical tone is due to a rapid periodic motion of the sonorous body; the sensation of a noise to non-periodic motion."[33] The complex irregular sound of noise overloads the listener's capability to understand sound, presenting a chaotic and unstable set of relationships engulfing the order and simplicity of pitched sound.

Noise, however, cannot simply be understood as solely an audible phenomenon. In information theory, noise is defined as an intrusion into the process of communication. Information theory was originally formulated in 1948 by Claude Shannon, a mathematician at Bell Labs, a division of the Bell Telephone company, in a paper entitled "A Mathematical Theory of Communication."[34] He developed the theory in an attempt to rid telephone communication of excessive noise. His theory uses mathematics to calculate how much information a given channel can carry and the ratio of signal to noise within that channel. The signal in the line is understood to be the message or what is being communicated and noise is anything extra to the intended message.

Thus in information theory noise is understood as anything extraneous to the message: this includes everything from pauses in dialogue (for example, "umms" and "errs"); to a smudged newspaper text; to even the interesting people at table nine who make your current conversation hard to follow.

Noise enters the channel of communication between the source and the receiver. In the case of music playback there are numerous points in the production chain at which noise can become present, for example a poor recording of a live event, or a scratch formed on a record's surface.

Clare Taylor in her paper "Noise Is OK" makes a distinction between noise simply created through the limitations of the media and accidental noise that is introduced further down the line of communication.[35] She gives the example of the medium of the newspaper and its inherent noise that exists in the drama of headlines, photographs, journalistic writing styles, layout, and advertisements. This noise is taken by us as being inherently related to the medium itself and has become almost completely accepted. The noise here is of such a constant nature that we can easily navigate our way through or around it.

Our much played favorite vinyl record is a good example of Taylor's distinction. Even though it has come to be filled with the noise inherent in the media—its many ticks and pops and the haze of ingrained dust—we continue to listen to it. In fact we forgive vinyl media for this flaw and even hear these noises with a sense of nostalgia, as they are marks created from listening to the record and remind us of times in the past when we played the music.

Paul Hegarty begins his book *Noise/Music* by bluntly stating that "Noise is not an objective fact."[36] For Hegarty the perception of noise occurs in relation to a historical, geographical, and culturally located subject, one whose listening

> is brought back to hearing through processes of rejection (as noise), confusion (through noise as change), excess (including of volume), wrongness or inappropriateness, failure (of noise, to be noise, to not be noise, to be music, not be sound, not be). Noise is where all this listening goes when it has had enough.[37]

Subjective noise is the most common understanding of what noise is. Put simply, it is the sound of the complaint from a stereotypical mother screaming to her teenage son to "turn that noise off." To the parent, the aggravating noise is the sound of the music, while it is his mother's voice that is noise to the teenager enjoying his music. Subjective noise is any sound that annoys, irritates, or hurts a particular listener. There are certain types of sound that more people will find to be noise, for example, high pitched sine tones, extremely loud sounds, brutal or aggressive music, and so on. Although this may be the simplest category of noise to understand—we all find certain types of sound or music to be noisy to us—it is also the hardest to set rules for.

By this definition music and noise are made distinct by the individual who separates the two by subjective and contextual means. If this is so, then noise can only exist in the ears of a subject who feels that the sound he or she is hearing is in fact noise. However, this becomes paradoxical when we consider the issue of noise music, a genre that is made up entirely of (acoustic) noise and yet is actually enjoyed by many. In addition, subjective noise does not account for the actualities of noise in terms of information theory, nor does it account for the acoustics of noise. Noise, in these cases, is not accounted for through the subjective experience the listener. That is, these noises are quite simply noise no matter what the subject feels about the sound he or she hears.

For Michel Serres, perhaps more than any other theorist of noise, noise forms the backdrop to all communication, the air we breathe and the sea from which all life emerges. His position takes in the areas charted by information theory, acoustic noise, and subjective noise, forming a type of materiality of noise:

> We breathe background noise, the taut and tenuous agitation at the bottom of the world, through all our pores and papillae, we collect within us the noise of organization, a hot flame and a dance of integers. . . . Background noise is the ground of our perception, absolutely uninterrupted, it is our perennial sustenance, the element of the software of all our logic. It is the residue and the cesspool of our messages. . . . Noise is the background of information, the material of that form.[38]

In *Genesis*, Serres writes about hearing, immersion, and background noise. It is the sea that is heard in *Genesis* as background noise. The sea is where for Serres we become most aware of background noise in our everyday lives. This unending noise of waves crashing on the shore is not a phenomenon as such but rather the opposite. As in acoustic noise theory, the sea, when understood as containing all possibilities, is where all life emerged and it is where all life will return. When something, information or music perhaps, comes into being it leaves noise, separated from its white chaos. Whereas information theory understands that noise is an impediment to communication, it is also a given that there will always be noise on the line; the key is abating it as much as possible. As Serres explains, noise is all around us and we are immersed in it constantly; in fact without it there can be no communication. It is the ground from which all communication is drawn and it is a constant in that communication.

Noise is embedded in information, in the object and subject of language; as the backdrop of communication, it fills the silences and gaps, often corrupting and confusing. Noise is repeatedly pushed to the background in an attempt to make it invisible and unknown. Whereas the background noise created by the ocean is all consuming—it cannot be ignored—other noises can easily be covered by the content of communication. Noise is not always loud: chaos can exist below the surface, a quiet backdrop to important rules and tidily ordered text and sounds.

In *The Parasite*, Serres equates repetition with death and theorizes that if the world fell into repetition all would be known: the extraordinary, the rare, and exceptional would be reduced to still, "flat" waters. But noise is never repetitious. In its pure form it cannot be known or expected; it is not logical.[39] Noise is filled with chaos and chance, filled with every possibility, and as a consequence it is impossible to divide and predict it.

Noise and Cracked Media

These four ways of understanding noise can be used to account for parts of the practice of broken and cracked media. Much of the sound produced through the utilization of playback technologies is, acoustically, noise. The rasp of a needle drawn across a piece of sandpaper is acoustic noise, as is the snap of a breaking vinyl disc. This quality on its own, however, does not account for the nature of numerous breaks or cracks that are then brought together to form a piece of music or a performance.

Many of the sounds are not adequately accounted for in the scheme of acoustic noise. For example, the loss of information that causes a CD player to glitch and skip is not acoustic noise. Rather, here is something much more akin to noise in terms of information theory. If the clean playback of a CD is the aim of playback and the optimum situation for communication, then the loss of data leading to a skip can be considered “noise on the line.” A small loss of information is completely covered by the error correction data stored on the disc, but a larger loss of data leads to a digital glitch causing the music to jump, skip, and hang. If a pop on a vinyl disc is minor enough then we can listen through this small loss of data, but a larger scratch will remove the needle from the groove, causing a screeching noise as the needle slides across the surface of the disc.

However, in the case of, for example, the broken and reconstructed records of Milan Knížák, we are faced with an extreme loss of information. These cracked and broken records do not merely carry small imperfections, they are covered in cracks—and yet they continue to be played. The phonographic medium’s inherent noise blasts over any signal originally intended by its recording artist. To put it bluntly, the sound and music created by the producers of cracked media is too excessive to be heard as merely noise on the line, well in excess of any disturbance to communication and well in excess of inherent noise.

The crux of the issue is that performance practices and the noise produced folds in on itself in the process of recording and the release of the works themselves. For example, Knížák produces music by cutting and re-forming old records, and the noisy sound of these records generates the sound for a performance. This performance is recorded and is itself released on a record. How do we discriminate between wanted noise and unwanted noise in a recording of music that involves noise as a key part of the composition?

If noise is both a distraction and an abstraction (that which is removed from communication),[40] then we have an obvious problem when thinking about noisy music and sound. The problem is of course that the noise, the interference, is a key part of the signal and the meaning of the work. The work is filled with noise, but this noise is not a distraction from the "real" meaning; it does not disturb or disrupt the flow of communication at all. If this is the case then there must be another layer of noise in the work: if all communication is affected by noise, what is the noise in this case? This argument quickly leads to a feedback loop: are we hearing noise or are we hearing the content of communication, are we hearing the noise of the noise of the content of communication?—and so on.

A possible way through the dilemma is that this noisy music, as Taylor suggests, makes "noise OK." That is, by becoming familiar with noise as music we are able to move beyond the initial desire to cover or remove noise. Continuing with the example of Knížák, once we hear his "Broken Music," we become attuned to the possibility of hearing what we are accustomed to hearing as noise as the actual content of music, and can thus listen to these sounds beyond this single work. Although this might account for the noisiness of much contemporary music, it also removes some of the power the music has. For example, as Nick Smith argues, if Japanese noise music is situated as an anticapitalist aesthetic—as music that cannot be owned or that is made beyond capitalist exchange through extreme sounds—then we have a problem when noise is deemed "OK." Japanese noise music would, as Smith argues, then lose its power and be subsumed back into the exchange system.

Cracked media can be heard as subjective noise in a number of ways. The sound of a skipping CD played at volume is an unwanted sound to most. However, to those who utilize these sounds in the production of their compositions, or to those attuned to this music as an audience, these sounds become aesthetically interesting, and even prized. As will be discussed in chapter 3, "Damaged Sound," San Francisco music producer Lesser actually enjoys and seeks out the sounds of skipping CDs. There are also those (including myself), for example, who enjoy hearing a CD skip in a café. In this case it is the shift from music as quiet background to annoying noise that is of interest.

The practice of cracked media can also be heard as deliberately playing on the expectations of an audience for music, unexpectedly throwing them into noise. This noise might well blast them out of the comfort of a safe musical performance. This tactic can only work a small number of times, however, before an audience comes to expect the blast of sound or the noisy destruction of musical instruments. Here too we can imagine an audience split between those who find the sounds to be noisy and shocking, and those who hear them simply as sounds and an expected part of the performance.

Paul Hegarty writes, "Noise also has to contain judgement: it is 'unwanted.' Can noise be wanted—clearly that would define the noise in question as not-noise."[41] Hegarty argues that if we are to listen to noise as music, or if music is deliberately created out of noise, then we need to look at its musicality differently. Noise in broken and cracked media also takes structural form. That is, it forms the composition, or even becomes the score for the work. The known, planned accident is combined with the unknown chaos of noise. Here noise is the framework or support of a deliberately indeterminate order.

As Jacques Attali states in *Noise: The Political Economy of Music*, "Nothing essentially happens in the absence of noise."[42] In a materialist approach to noise, he argues that everything happens in the presence of noise, and through music, which he calls "the organization of noise," we can come to better understand the nature of our society and culture.[43] Looking to information theory, Attali takes an altogether different approach from Serres, arguing that "Noise . . . does not exist in itself, but only in relation to the system within which it is inscribed. . . . Long before it was given theoretical expression, noise had always been experienced as destruction, disorder, dirt, pollution, an aggression against the code-structuring messages."[44] For Attali, noise creates meaning through the interruption of the message, and through the freed imagination of the listener within pure noise, "The absence of meaning is in this case the presence of all meanings."[45] For new meanings to be created a crisis or catastrophe must occur, or perhaps an accident, that will focus the elements of chaos into a singular focused emergent meaning. Noise is then filled with all future possibilities.

If noise is an inherent part of the systems of communication and carries all possible futures or outcomes, then the tools of mediation also inherently hold all possibilities. The practitioners of cracked media take the objects of recording and playback and generate new outcomes for them utilizing noise—a noise that is always part of the system, waiting in the background. These artists generate unique and singular sounds and performance practices out of the chaos and disorder of noise, exploiting the possibilities present in the slightest hum on the line, or pop and crackle.

Noise has been brought forward to become the actual content of much recent work in the area of cracked media. However, this practice need not be seen as transgressive, excessive, or even chaotic, contrary to much of the theoretical discourse around it. Although at times noise utilized in this way certainly is excessive, it is just as often quiet, gentle, low, and almost inaudible. Noise need not be seen as disturbance; it need not be excess or transgression. Noise is the backdrop to all communication, but in those instances when the backdrop is brought to the fore it is simply not disturbing or blotting out any information; it is not a break in communication, but instead becomes the content of communication itself.

The next two chapters will look closely at the history of cracked media, teasing out its multiplicity of approaches. In the end we will see an extremely productive tactic in the twentieth century, one that is attuned to the expansion of sound and closely related to movements across the arts.

2 broken music

the manipulated, modified, and destroyed phonograph

1 Broken Music and Early Phonographic Experimentation

Numerous musicians, composers, and artists approached the phonograph throughout the twentieth century not simply as a mediating device for the playback of music, but as a device for the generation of unique sounds, indeterminate compositions, and art objects. The development of the turntable as a production device and as an instrument in itself has not always been carried out gently; performers have at times turned to destructive techniques, laying waste to countless records and destroyed phonographs. By the end of the twentieth century the turntable had been modified by experimental musicians to the extent that it was no longer capable of playing records back; in fact in much of the turntable practice of the late 1990s, records were not used at all. The techniques covered in this section range from the most minor manipulation—such as slowing down the speed at which the record revolves—to outright destruction—the smashing or burning of records.

The often destructive approach to the use of the phonograph has the effect of further filling the sound world with noise. The sounds produced by these musicians were filled with ticks and pops as well as rasping, grainy noise, often (but not always) played at high volume in performance. For many producers this approach was also closely related to indeterminate compositional practices. The sound outcome of a stylus hitting a large scratch in a record has a predictable popping effect. However, it is not known how the tone arm and stylus will react to the physical collision with the scratch. Will it, for example, simply hit the scratch and carry on playing, or will the indentation in the vinyl cause the needle to jump its groove?

The Phonograph as Instrument

The phonograph,[1] though designed as a technology for playback, has also been widely used an instrument throughout the twentieth century. In this practice the technology shifts from simply reciting the content stored on the vinyl disc to creating content in both live and studio situations. The technology has been used to create original and additional sounds, both in live performance and the studio context, and has often been used in an instrumental manner. Christian Marclay made this explicit when he performed with a customized turntable, called the "phono-guitar," in a guitar-like

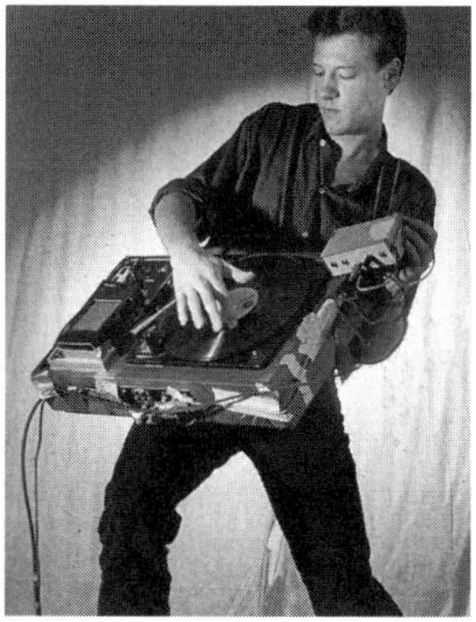

2.1
Christian Marclay playing the Phonoguitar, 1982. Photograph: Steven Gross.

manner; similarly, but later, in the music video for Tone-Loc's "Wild Thing" (1989) the DJ uses a turntable slung over his shoulder like a guitar.[2]

The turntable is ubiquitous in hip-hop and DJ culture. Hip-hop, in its desire to be understood as a legitimate musical form, has been at pains to establish the concept that the phonograph is an instrument. This sentiment appears in a number of texts and is put most succinctly by Charles Mudede in "The Turntable," where he simply states, "The turntable is an instrument."[3] That the turntable is used as an instrument is without question for the hip-hop turntablist or the dance DJ. The DJ under these conditions is understood to be a musician, and his or her instrument is the turntable. The popular music practices of hip-hop and DJ culture do not, however, account for the practices under discussion here. Instead we need to look deeper into the history of extended techniques and prepared instruments.

Prepared and Extended

Extended techniques and prepared instruments are a key part of the expanded sound world of twentieth-century musical practices. These practices are not at all unusual or novel in that there are countless examples of the practices in numerous different musical cultures. As such the discussion below provides only a few notable examples.

Extended techniques are techniques for playing instruments that are beyond the traditional or expected manner of performance. This could include, for example, playing a violin with the wooden part of the bow or bowing the body of the instrument itself. Prepared instruments are a subset of extended techniques, in that preparations involve the usually temporary modification of the instrument to alter and extend the sound outcomes. Gary Butler dates the first experiment with a prepared instrument to a performance of Erik Satie's *Le Piege de Méduse* in 1914.[4] Satie inserted paper between the strings of his piano, thereby extending the possible sounds the piano could make beyond its traditional use. From this first seemingly impulsive action have come numerous uses of the technique both in Western art music and in other genres such as rock, improvisation, and experimental musics. The best-known practice of the prepared instrument is that of John Cage in his compositions for prepared piano. Cage originally affixed screws, nuts, and bolts to the strings of an upright piano and later developed compositions and scores around the preparation of grand pianos.[5] The placing of these objects inside the piano caused a change of timbre and extended the range of possible sound the piano could be expected to create, thereby expanding the sound field of the instrument.

Extensions and preparations are not just an element of Western art music; they have become a common element of contemporary improvisation and are now an expected element of performance in improvisational practice. For example, objects such as drumsticks, rulers, bits of metal, wire, springs, foam, cloth, and pegs are often wedged between the fret board and strings of guitars or double basses. Once these objects are inserted, musicians, such as Keith Rowe or Sugimoto Taku, might gently pluck, move, or hit these objects, thereby causing the strings of the guitar to resonate in a manner not at all typical of the instrument. Keith Rowe also uses motorized objects placed on or near the strings; the vibrations are picked up by the strings and the guitar's pick-up device and are subsequently amplified. A similar approach has also been taken to percussion; musicians might place foreign objects on the drum skins, activating the drums via various manipulations of an array of detritus. The Australian percussionists Tony Buck and Robbie Avenaim place chains, rattles, sticks, stones, motors, and metal discs on their drums. During performance these objects are manipulated and activated by hand or hit with drumsticks or other objects. As one might imagine these techniques greatly extend the sound outcome of the instrument and remove it from the confines of traditional musics.

Another approach to preparation is the incorrect wiring or plugging in of electronic devices so as to deliberately cause feedback loops. Musicians such as Nakamura Toshimaru wire their mixing desk incorrectly by plugging the outputs into the inputs and vice versa. In performance his desk produces various high-pitched sine-tone-like sounds, and he also makes various beating sounds as he finely manipulates the instrument.

John Cage's prepared pianos have had a direct influence on the expanded sound world of rock music. The prepared and extended instrument is not solely the domain of Western art music, or of experimental music, as elements of rock have made use of both preparations and extensions. John Cale, who played in the Velvet Underground, for example, cut down his viola's bridge so as to enable all the strings to be played at once, rather than the regular maximum of two. He also at times prepared his viola with electric guitar strings.

The band Sonic Youth has continuously developed extended techniques and preparations for their guitars over the last few decades. The band, since being formed in 1981, has made use of some 120 guitars that were prepared and modified in various ways, which they then played often using extended techniques.[6] Preparations include numerous different tunings, as well as the modification of the guitars themselves. In performance members of Sonic Youth regularly place drumsticks between the strings of their guitars. These sticks are struck causing the strings to resonate in different ways depending on where they are placed along the fret board or in relation to the pickups. The drumsticks are also rubbed across the strings in a more aggressive version of the use of a regular slide.

What connects these varying approaches to prepared instruments and expanded techniques is an interest in variability. These musicians manipulate instruments to expand their sonic palette, thereby varying the expected sound outcomes the instrument is capable of. By customizing these instruments, and customizing their performance techniques, these musicians show a shared interest in novel and unique sounds and an expanded sound palette.

The Extension and Preparation of the Phonograph

"The scratch" is perhaps the most well-known practice of turntable extension. In its most simple form it is the technique whereby the DJ moves a record backward and forward while the needle is sitting on the disc's surface. This action commonly occurs swiftly and may render the sounds from the original recording all but unrecognizable. In addition to this basic technique, numerous "trick" techniques have been developed that extend the audio outcomes of the scratch and create a virtuosity of the scratch.

The scratch is a form of turntable manipulation and if done correctly causes no permanent damage to the record. As will be shown here, however, this type of manipulation leads to more severe and damaging techniques. Hip-hop is often regarded as the initiator of the scratch, and musicians using turntables outside hip-hop, such as Christian Marclay, are perceived as simply developing more extreme versions of the art of hip-hop turntablism.[7] However, as will be discussed below, this practice has far earlier precedents. Record manipulation occurred early in the twenieth century as the simple inquisitive act of slowing the record down, playing it backward, then playing it backward-and-forward. In terms of a historical progression, the DJ scratch came quite some time after early techniques for phonographic manipulation, crack, and break, but because of the mainstream nature of hip-hop and DJ culture their techniques brought the use of the turntable as an instrument into focus and influenced or caused later experimental turntablists to react against their techniques.

The preparation of instruments is usually done with great care so as not to damage the instrument, allowing it to return to its previous state. The cracked or broken preparation, however, often causes permanent damage and even destruction. The effect of these modifications is to deliberately generate sounds not originally produced as part of the recording or to generate sounds not intended to be produced by the phonograph by modifying the device itself. In the most extreme form of such modifications, the salvaging of the original recording is not possible, leading to the permanent silencing of the content of the recording. Once the record has been damaged by a preparation, by having its surface deliberately scratched or having indentations burnt into it, for example, the record is ruined for its original purpose. This causes the object to be more susceptible to further preparations and destruction, as it does not hold its original worth as an object. That is, the devaluation of the record as a functional commodity, and as a fetishized object, leads in the end to its complete destruction. What begins as an annoying scratch, which causes a rhythmic pop on playback, is pushed further by deliberately gashing the surface, causing the needle to jump grooves when it is played.

Numerous approaches to cracked and broken phonographic technologies have occurred over the years since its inception in the early twentieth century. The approaches can be broken down into the three basic areas of manipulation, cracking, and breaking. In terms of the use of records we can classify these practices as follows:

1. Prepared records—cracked
   a. Temporary preparation of the surface of the record disc. For example, placing removable tape on the surface, which causes the needle to slide across grooves.
   b. Permanent preparation of the record disc, leading to permanent damage. The disc is cracked but is still playable. For example, physically scratching the surface of the disc to produce pops or to force the needle to jump grooves; cutting the grooves to form locked grooves; painting the disc and filling grooves with paint or other substances such as mud; deliberately warping the record so it is no longer flat; heavily scuffing the surface of the disc with sandpaper.

2. Prepared records—broken
   a. The record is "prepared" and re-formed to produce a somewhat playable disc. For example, cutting the disc into wedges and re-forming the disc.
   b. The record is completely destroyed and is beyond playback. For example, smashing or burning the record.

The experimental turntablist also manipulates the phonograph itself through various extended and prepared techniques, which include:

1. Manipulation of the phonograph
   a. Sound is produced via the nondestructive manipulation of the phonograph. For example, forcing the plate to play backward or at the wrong speed; or allowing the spring mechanism of a gramophone to run down.
   b. Sound is produced through the possibly destructive manipulation of the phonograph. For example, playing the phonograph without the record disc and placing the needle directly on the turntable's platter.
   c. Sound is produced through the manipulation of the phonograph with external objects. For example, the

turntable is used as a rotary device, and objects or microphones are bumped as the deck of the phonograph turns; or a cymbal is held over a turntable's cartridge head and causes a feedback loop to occur.

2. Preparation—cracked
   a. Sound is produced via the preparation of the phonograph with objects not intended for playback. For example, replacing the vinyl disc with other objects such as sandpaper, wood, or metal.
   b. Sound is created through the permanent modification of the technology. For example, by replacing the stylus with a toothpick or with contact microphones.
3. Preparation—broken
   a. Sound is created through the permanent and destructive modification of the technology. For example, the tone arm is removed from the turntable deck or the cartridge head is removed from the tone arm altogether.
   b. Sound is produced via the destruction of the turntable. For example, smashing the tone arm down on the platter of the turntable.

The Trajectory of the Cracked and Broken Phonograph

In general, the crack and break in relation to the phonograph has followed the technology's changing value as a commodity. In its early life the phonograph was treated with care; in these practices, it was simply manipulated. Records themselves were 78 rpm shellac discs, sturdier than the thin, mass-produced vinyl that became prevalent in the 1960s. The original technology could therefore be handled much more physically, and could withstand a greater amount of abuse before breaking. Phonographic objects were more robust, but they also held more value as a commodity; their destruction was undesirable for the owner and unexpected for the audience.

In the middle years of the twentieth century, phonographic technology became an everyday technology; it was taken into homes as the main musical playback device. As the technology became more prevalent and accessible, it also became available for experimentation and accessible for use in performance. An example of its use in the live context is the early work of Nam June Paik, who often played records during his performance events in the late 1950s. The everyday nature of the phonograph resonated with the everyday approach to art taken by the Fluxus group, though its use soon turned to abuse. This moment is a key point in the development of an experimental use of the phonograph and the first example of the destructive use of the technology.

Fluxus was, however, a discontinuity in the development of the crack and break: it broke with the trajectory, jumping ahead to what would not subsequently reoccur for some thirty years. The reasons for this will be discussed in the section below entitled "The Crack and Break in Phonographic Sound Production."

By the 1970s the record itself had changed in value with the accumulation of a huge stockpile of cheap, mass-produced vinyl records. The turntable itself, however, especially in hi-fi aficionado circles, actually increased in value. This situation allowed for the vinyl experimentation to develop in advance of cracked and broken turntable experimentation. However, since the 1990s the musician using records has had a different relationship to the phonograph from those working previously. Mass culture has left the record itself behind, clearing it out of the mainstream market's music stores and production lines. The record now exists predominantly for use by the DJ or in the careful hands of the collector of esoteric original recordings and hi-fi sound playback. The effect of the obsolescence of the phonograph has been to open it up to the possibilities of extreme preparations and modifications, as both the records and the record player have little or no value in the current music market.

The sound of the scratched record or needle jump caused by damage to the surface of the record was, as Thomas Levin points out, “an acoustic event that only recently was still disturbing enough to require one to interrupt a recording before it had run its course.”[8] For Levin the scratched record has become a nostalgic trace of a bygone era owing to the development of digital audio. Analog malfunctions that cause a skip or pop on a scratched record have thus become available as aesthetic practices, as the possibilities of the scratch and skip are no longer connected to the failure of the technology or irreparable damage to the vinyl object.

The Prehistory of Cracked Media: Early Experimentation with the Phonograph

> There is only one point at which the gramophone interferes with both the work and the interpretation. This occurs when the mechanical spring wears out. At this point the sound drops in chromatic weakness and the music bleakly plays itself out. Only when gramophonic reproduction breaks down are the objects transformed. Or else one removes the records and lets the spring run out in the dark.
>
> —Theodor Adorno[9]

The quote above comes from an essay (originally published in 1927) written by Theodor Adorno when he was only twenty-four. The essay centers on Adorno's concern for the future of listening in the face of phonographic technology and can be related to later issues of contention between Walter Benjamin and Adorno concerning Benjamin's "The Work of Art in the Age of Mechanical Reproduction."[10] Adorno believed that the future of close listening was in jeopardy, as people were taking an increasingly imagistic approach to listening as a result of the prevalence of record technology. No longer was the audience firmly seated in the concert hall, paying full attention to the piece of music for its full duration. Instead the audience was in the comfortable and distraction-filled space of their own home. The above quote, though, is an anomaly in the essay. The idea that the gramophone could transform sound via its own mechanical breakdown is not alluded to at any point until the final paragraph of the writing.

This quote suggests the possibility of the phonograph actually transforming objects. The recorded disc creates something new as the equipment "breaks down." Although it is a natural part of the original mechanics of the phonograph to wind itself down while playing, it is this effect that Adorno here singles out. He hears this moment as the only one of possible transformation. It is this point of breakdown that this chapter is most interested in: the moment when the phonograph winds down, the disc plays slowly, or the inquisitive listener manipulates the record to play it backward. We hear in Adorno that it is the break that is important, the transformation of the intended sound and playback. In the prehistory of the crack and break composers focused on the manipulation of the technology, and it was not until the 1960s that the phonograph started to be physically altered.

The Invention of Recording Technology

At the time of the phonograph's invention in 1877 Thomas Edison had various ideas about its possible uses, including dictation, phonographic books, teaching elocution, reproduction of music, recording messages from family (including one's dying words), music boxes, articulate clocks, the preservation of languages, an educational tool, and various uses in connection with the telephone.[11] The phonograph was developed steadily over the next seventy years, including the replacement of the cylinder with flat shellac discs allowing for multiple copies of recordings to be easily reproduced, the development of the vacuum tube, and electronic amplification and recording.

It was not, however, until the 1920s and early 1930s that the phonograph began to be actively experimented with as a source of original sound production, rather than the means of playback of a recorded piece of sound. Composers in the 1920s and 1930s, such as John Cage, Percy Grainger, Paul Hindemith, Raymond Lyon, Darius Milhaud, László Moholy-Nagy, Ernest Toch, and Edgard Varèse, were interested in such creative applications. These phonographic practices have been excellently documented by Richard Schmidt James in "Expansion of Sound Resources in France, 1913–1940, and Its Relationship to Electronic Music," and by Mark Katz in *The Phonographic Effect: The Influence of Recording on Listener, Performer, Composer, 1900–1940*.[12]

Joel Chadabe observed in *Electric Sound* that a major issue with the use of the phonograph for original work in the 1920s and 30s was that it did not allow for editing sound.[13] The problem of the fixed nature of the disc was overcome in the late 1940s by the composers of the musique concrète school, who used multiple turntables (and much later in the late 1970s and early 1980s by DJs who deftly took sound bites from records and mixed them with others using a DJ mixer and two or three turntables). However, a small number of composers did make works using the phonograph as the central instrument in itself and a few writers (discussed below) described various possibilities for the future of music and sound expansion through the extended uses of the phonograph to produce new sounds and methods for composition.

During the 1920s and 1930s the phonograph was the main source of sound reproduction and playback technology available. The wire recorder was not available until 1929, and the first magnetic tape recorder was completed in 1935.[14] In the early 1930s, according to Richard Schmidt James, the relative merits and pitfalls of the technology were debated along with its future. Raymond Lyon suggested that since the record was inscribed by humans, they could "impose and direct deformations of phonographic reproduction," and that therefore the equipment was capable of more than the mere reproduction of sound.[15] James states that Lyon described several means of deforming phonographic reproduction, including changing the speed of the turntable's rotation, the complete relaxation of the driving spring, and the juxtaposition and superimposition of multiple recordings. Lyon is quoted as stating:

> Can you imagine, then, what might be done by manipulating the *recording process*? . . . One might conceive of the existence of an esthetic of the phonograph which permits

> the knowing instigation of deformations intended to produce certain effects.[16]

Lyon is suggesting a future method for the use of the phonograph as a compositional tool with its own aesthetic. The phonograph is described not merely as a means for reproduction, but as a possible means for the creation of new sounds in future compositions. Phonographic recording techniques that change the revolution speed during recording, as well as use the techniques of cinema editing, are, for Lyon, the way forward for the technology. In addition he suggests that nonmusical noises and sounds could be recorded and manipulated and gives the example of a revving car, the slamming of a door, and the whistling of a paper delivery boy.[17]

By the time Lyon's piece was printed in 1930 there had been a number of experiments in exactly these sorts of techniques. Arthur Hoérée recalled electricians experimenting in the 1920s with varying the speed of the turntable with their fingers, then letting the spring unwind to allow the disc to be played backward.[18] Composer Darius Milhaud had been experimenting with vocal sounds by manipulating the rotation speed of the disc as early as 1922.[19]

Igor Stravinsky, who had a long-standing fascination with machinery, saw the potential of the phonograph as an archival tool as well as a direct model for how the composer wished the piece to be played. He had been a record collector since 1914, collecting American popular music.[20] His *Serenade in A Major* was composed in 1925 expressly for recording. This was not the very first example of this approach, but it was certainly an early model for the future of the phonograph. Stravinsky was, however, the first composer to undertake to record all of his works.[21] In 1930 he called for recording itself to take advantage of the plastic capability of phonographic reproduction. He explains:

> It would, however, be of the greatest interest to create music specifically for the phonograph, music whose true image—its original sound—could only be preserved through mechanical reproduction. This would well be the ultimate goal for the phonographic composers of the future.[22]

The true form of the music, "its original sound," would thus only come about in the playback of the record.[23] This position foreshadows that taken by Fluxus composer Yasunao Tone in the early 1990s and points the way for the record to become more than merely an imperfect reproduction of an actual event.[24]

In a discussion of the generative, rather than the reproductive capiblities of the phonograph, László Moholy-Nagy opens his article, "New Plasticism in Music: Possibilities of the Gramophone," by mentioning musical experiments with amplifying tubes and their capacity to generate new sounds. Moholy-Nagy suggests that the record could be etched into directly by the hand of the composer. One could not inscribe these marks accurately onto a regularly sized disc and as such a master that was very large would need to be created, this would later be drastically reduced.[25]

Moholy-Nagy was not the first to put forward this idea. Alexander Dillman in 1910 suggested that one day it might be possible for an engraver to engrave sound waves directly on the surface of the disc. He was excited by the idea that the once abstract, ephemeral form of the voice would be physically etched onto the surface of the record and could be seen with the use of the microscope. He also imagined that if we could reproduce these sound waves by mechanical means we could create a singer with unlimited range.[26]

Moholy-Nagy's call for the composer to be able to produce music without the aid of a performer reproducing the work was echoed in 1937 by John Cage in his influential paper "The Future of Music: Credo," in which he states, "It is now possible for composers to make music directly, without the assistance of intermediary performers."[27] Cage goes on to argue that centers for electronic music need to be set up, and within them "oscillators, turntables, generators, means for amplifying small sounds, film, phonographs etc." will be available for use.[28]

Theodor Adorno wrote in 1930 about the place of the phonographic recording in musical culture but denies the possibility of a gramophone-specific music. As already shown, however, by 1930 gramophone-specific music was already well on its way. In his article "For the Record: Adorno on Music in the Age of Its Technological Reproducibility," Thomas Levin states that Paul Hindemith and Ernst Toch had produced an event called *Neue Musik Berlin 1930*. This work was mentioned in the German Music journal *Melos*, which Adorno also wrote for, by Heinrich Burkhard:

> This made-for-phonograph-record-music was accomplished by superimposing various phonographic recordings and live musical performances, by employing variations in speed, pitch height and acoustic timbre . . . the result was

> an original music which can only be recreated by means of gramophone apparatus.[29]

Three 78-rpm discs are known to have existed of Hindemith's experiments, although these discs were given to a junk dealer when Berlin's State Institute of Music Research was not interested in them.[30] Mark Katz tells us that Hindemith's first public performance of phonographic music was held on June 18, 1930, in Berlin. The performance utilized three one-sided ten-inch discs containing singing and instrumentals that were played at different speeds.[31] Katz also believes that the recording itself was mixed with multiple phonographs, each of which plays one instrument into the final recording.

Toch also had a piece, *Gesprochene Musik*, performed at the same concert. He explained his use of the phonograph in the journal *Melos*:

> the concept arose from the attempt to extend the function of the machine . . . by exploiting the peculiarities of its function and by analyzing its formally unrealized possibilities . . . thereby changing the machine's function and creating a characteristic music of its own.[32]

Toch used voice as the instrument of these experiments and changed the pitch of the voices by changing the speed of the phonograph during playback. He tells us that the result is an instrumental music not recognizably from a vocal source. As Levin writes, this moment of phonographic experimentation is a key point in the "archaeology of the contemporary practices of gramophone-specific music referred to as 'scratch' and 'mix.'"[33] Interestingly, neither Toch nor Hindemith experimented further with the phonograph as an instrument.

In 1939, Edgard Varèse, with the aid of an engineer, experimented with creating electronic music using a variable-speed phonograph.[34] Although Varèse did not continue in this direction, he saw his experiments not as music but simply as examinations in sound creation made in advance of any musical fact.[35] Neither the specific use of the phonograph nor the sounds it created were included in any of his later compositions, although they no doubt had an influence on later tape pieces such as *Poème Électronique* (1958).[36]

John Cage: From Manipulated Turntables to Cracked Media

It would be John Cage who first saw, with the greatest clarity, the possibility for the phonograph as an instrument and who went a long way toward the cracking of the medium. Cage over the course of his career had a complicated relationship with the recording medium and saw its potential for the composer not in the recording and archiving of performance, but rather in the possibility for new and extended techniques and practices. He argued:

> though people think they can use records as music, what they have to finally understand, is, that they have to use them as *records* . . . the only lively thing that will happen with a record, is, if somehow you would use it to make something which it isn't. If you could for instance make another piece of music with a record.[37]

Cage was in the audience in Berlin for the 1930 performances of Toch and Hindemith. He is said to have commented that Toch was "onto some good stuff."[38] Cage himself experimented with the phonograph in 1939, using two turntables in his composition *Imaginary Landscape No. 1*. He was studying the application of electronic technology in music at the Cornish School radio station when he discovered variable-speed turntables. The decks included a clutch to shift between speeds, making them ideal as a possible instrument. In 1939 Cage was asked to compose a piece for a Seattle performance of *Marriage at the Eiffel Tower* by Jean Cocteau. The piece Cage wrote was for muted piano, cymbal, and two variable-speed turntables. One of the turntables played a Victor frequency record, 84522B, and a constant-note record, No. 24, and the second turntable played the Victor frequency record 84522A.[39] These test records were produced to test the frequency response of the phonograph and were not intended to be played as music. The piece was performed by Cage, Xenia Cage, Doris Dennison, and Margaret Jansen in two radio studios before being mixed and broadcast a few steps distance from the theater (the work was composed to be performed as a broadcast of a recording). The piece was premiered on March 24, 1939, and later used at Mills College for Marion van Tuyl's dance *Horror Dream*.[40]

When the record speed was changed the test-tone records changed in pitch; their utilization in this piece was one of the first compositions to use "electronic instruments." The piece was described by Virgil Thomson, in 1960, as suggesting:

> a sorcerer (or his apprentice) in a radio studio, where electronically produced sounds can swoop around and

> fly through air with the greatest of ease. The sounds lack charm (as all electronically produced sounds do), and technologically the work is primitive. But for its time, ten years before tape, it is remarkable.[41]

Imaginary Landscape No. 2 (1942) also used turntable equipment, but in this case Cage modified the turntable. The piece is for five musicians who play tin cans, conch shell, ratchet, bass drum, buzzers, water gong, metal wastebasket, and lion's roar. The work also calls for a phonograph tone arm with the stylus replaced by a coil of wire, a technique Cage learned from a radio sound effects producer.[42] Cage explains:

> when I got to Chicago, I had a commission to do a piece for CBS and I worked with a sound effects engineer at the radio station in Chicago and he showed me the thunderous sound of the coil wire in a contact microphone—which I loved.[43]

Imaginary Landscape No. 3 calls for a battery-powered buzzer, an audio-frequency oscillator, a marimba with contact microphones, and three variable-speed turntables—two played with constant frequency recordings and the third with a recording of a generator whine and a cartridge with a radio coil. The piece premiered at the Chicago Arts Club on March 3, 1942, and was performed by Xenia Cage, Katherine Manning, Brabazon Lindsey, Marjorie Parkin, and Stuart Lloyd.[44]

Cage was to take his experiments one step further, focusing on the prepared cartridge head detached from the phonograph's deck. He developed this approach into a composition that appeared a few years before the musicians of Fluxus would mount a more destructive assault on musical instruments, including the phonograph.

2 The Crack and Break in Phonographic Sound Production

Cage's Modified Phonographs

For the highly influential composition *Cartridge Music* (1960), John Cage began to further modify the phonograph's cartridge head. As the title suggests, the work focuses solely on cracked phonograph cartridges. For this piece the cartridge was modified by removing the needle and replacing it with foreign objects, including pipe cleaners, wires, feathers, slinkies, matches, toothpicks, nails, twigs, cocktail parasols, and miniature American flags.[45] Sounds within the work are produced by striking or rubbing various objects with the cartridge.

The sounds are picked up by the device, amplified, and then played back through speakers placed around the room.

The score for *Cartridge Music* consists of twenty transparent sheets, each showing one to twenty irregular hand-drawn shapes. The number of cartridges employed is matched to the number of shapes on the transparency. This sheet is then overlaid with four transparencies marked with points, circles, a curved line, and a circle marked with a stopwatch. The intersections between the shapes mark moments in the performance and direct the performer as to how to use his or her cartridge—for example, whether to hit or scrape the cartridge over an object—as well as directing the level and tone of amplification. Cage said of the piece:

> The sounds that result are noises, some complex, others extremely simple, such as amplified feedback, loud-speaker hum, etc. (All sounds, even those ordinarily thought to be undesirable, are accepted in this music.)[46]

This music uses the phonograph and the resulting transformations to produce the content and sound of the work and embraces the notion that all sounds are available as the material of music. The cartridges have in a sense been ruined, the diamond needle replaced with a slinky or a bit of stick. The record has been replaced with objects and various pieces of furniture, over which the cartridges are drawn. Hi-fi stereo sound has been replaced with scrapes and scratches, speaker hum, sometimes loud, sometimes inaudible. The phonograph is cracked, but not quite broken or destroyed, as it still functions to produce the noises and audio desired by the composer.[47]

Cartridge Music marks the beginning of cracked media in relation to the phonograph. It is a fully realized composition for which the modified turntable is the only instrument. As the previous section has shown, the phonograph had developed from reproducing a previously recorded piece into an instrument able to make its own sounds and even possess its own aesthetic. There is something subtly violent in the near-vandalism of the technology introduced by Cage, as the technology is used to produce sound outside the parameters of conventional hi-fi playback.

In 1960, when *Cartridge Music* was composed, the sound and art world were shifting. It was at this time that proto-Fluxus events and early Happenings began to occur. Cage may have pulled the needle from his cartridge, but very soon both the turntable and the record were to go through some extreme abuse and destruction in the name of sound expansion and new processes for creating music. The two subsequent composers who took to mediating technologies most proficiently, and at points most destructively, where Nam June Paik and Milan Knížák. Both these artists are closely associated with Fluxus, a movement that was in turn heavily influenced by Cage. Because of these connections it will be important to gain an understanding of the approach Fluxus took to music and to the destruction of instruments. It will be seen that these approaches are key parts of the practice of both Paik and Knížák and also heavily influence later practitioners of the crack and break.

Fluxus, Music, and the Phonograph

> There are legends on top of legends surrounding Fluxus, some true, some half-true, some so ridiculous they could never be true and some so good they ought to be true.
> —Ken Friedman and James Lewes[48]

Fluxus work is often confusing, messy, loud, difficult, funny, ridiculous, serious, conceptual, simple. It is in these various contradictions and oppositions that the loose movement can be understood to have produced it most long-lasting and influential work. Fluxus was an interdisciplinary movement that included performance, theater, sculpture, printmaking, installation, design, poetry, mail art, and music. It is the area of music that will form the focus of this section. It should be understood, however, that music in Fluxus happens alongside the other aspects of the movement, and in many ways to draw music out goes against the intentions of the artists. Nonetheless, music is such a central element in the movement that it is useful to focus on the performance of music and look more closely at how music was approached by Fluxus. The other important element of Fluxus to explore is its approach to destruction. Music and destruction, when brought together, led to a rapid and early expansion of the use of cracked media specifically in relation to the phonograph.

Douglas Kahn observes in his essay "The Latest: Fluxus and Music" that Fluxus could count many of its members as composers and performers (in the musical sense), including La Monte Young, Nam June Paik, and Benjamin Patterson.[49] In addition, we can include others closely and not so closely related with the Fluxus group such as Philip Corner, John Cale, Henry Flynt, Ichiyanagi Toshi, Milan Knížák, Alison Knowles, Arthur Køpcke, George Maciunas, Yoko Ono, Marian Zazeela, and the members of Tokyo

Fluxus: Yasunao Tone, Kosugi Takehisa, Shiomi Mieko, and Mizuno Shūkō.[50] Many of the events most often counted as key proto-Fluxus events are in fact musical, including Cage's classes at the New School of Social Research between 1957 and 1959 and the series of performances that began in May 1961 held by La Monte Young and Yoko Ono in her loft apartment in the pre-Soho district. The loft series comprised work by Philip Corner, La Monte Young, Terry Jennings, Ichiyanagi Toshi, Henry Flynt, and Richard Maxfield. Another important proto-Fluxus series was the Darmstadt lectures and performances held in Germany by Karlheinz Stockhausen in the atelier of his wife Mary Bauermeister. These concerts included work by Stockhausen, along with Nam June Paik, Benjamin Patterson, John Cage, Ichiyanagi Toshi, George Brecht, and La Monte Young.

The relationship between Fluxus and music cannot be underestimated, and it is through this relationship that we will make the link between Fluxus and cracked media. Of particular interest is the work of Milan Knížák, Arthur Køpcke, Nam June Paik, and Shiomi Mieko, all of whom used turntables as elements of their work; Philip Corner for his gentle destructive compositions; and Yasunao Tone, whose Tokyo Fluxus work was highly influential in his development as a composer and who later used "wounded" CDs in his compositions.

Music Beyond the Inaudible

Some of the most subtle and sometimes perplexing approaches to music came from Tokyo Fluxus and especially the group Group Ongaku (Mizuno Shūkō, Kosugi Takehisa, Yasunao Tone, and Shiomi Mieko). The group heard sound as an object in itself, and the action, rather than the sound outcome, was privileged.[51] Shiomi, for example, at one group practice became sick of the "loud and rich sounds" being produced and decided to throw her keys in the air. She saw this as an action rather than as playing the keys like an instrument.[52] The difference between the action of throwing keys and the playing of keys is that the latter is aimed at producing a certain sound. Shiomi is not looking for a sound; instead she is simply throwing keys in the air, and the sound produced is of no interest. In fact it does not mater if a sound is produced at all.

Shiomi's *Boundary Music* (1963) asks the performer to make the faintest possible sound using instruments, human bodies, electronic apparatuses, and anything else.[53] During the years leading up to the work, Shiomi had defined one of the essentials of music as a "concentrated duration of activity involving the occurrence of sounds and silence."[54] She heard the simple act of throwing keys in the air as a task suitable for the creation of music and had moved on to create a piece in which the faintest and quite possibly inaudible sounds were created.

A more directly relevant example is Shiomi's use of the phonograph in performance. She used record players in a number of pieces, the most well-known being *Water Music* (1965). During one performance of the work at Gallery Crystal in 1965, Shiomi played a record of Carl Maria von Weber's *An Invitation to Waltz* that had been covered in soluble glue. When water was dropped on it the glue dissolved, exposing the grooves of the record underneath. The record was played during the performance and would have created an interesting juxtaposition of snippets of music played against the noise of the needle being drawn across the glue. Water was added until the entire record was exposed.[55] In a more destructive fashion, Shiomi's *Amplified Dream No. 1* (1969) used a recording produced by playing a phonographic record with random indentations caused with a soldering iron.[56]

Shiomi's work resonates with *Music While You Work* (1962) by Arthur Køpcke, who also uses blockages on the disc, but in his work these blockages direct the performance. The score for the work requires either scotch tape or glue to be placed on the surface of a record. The performer carries out a cleaning task such as sweeping the floor. When the needle sticks or is locked into a groove the performer stops what they are doing and resets the needle back to the first groove. The performance is over when the work is complete or if the record finishes.[57] Although numerous Fluxus works utilize cleaning as an activity, this work particularly engages with issues of mechanized and industrialized labor. Here the cleaner works when the phonographic machines tells him or her to, and as such the music both directs and is the background to the labor.

The composers of Fluxus had an extremely open definition of what could be considered music. They heard music in silence, in the beating of a butterfly's wings and even in the act of passing by a tree. Of course not all music produced by the composers of Fluxus was without audible outcomes, and numerous members took with equal zest to the expansion of audible sound through the techniques of prepared instruments and expanded techniques, which in turn gave way to all manner of cracked and broken instruments and playback equipment being included in their musical production. Fluxus took to instruments and mediating devices, tearing them apart, smashing them with hammers and hacking into them with saws. Instruments were set upon by the musicians who extended them to the point of breakdown and the complete destruction of their musical tools.

To understand how the cracked and broken media practices of Milan Knížák or Nam June Paik are to be heard as productive musical directions, it is important to look at this destructive turn and particularly the Destruction in Art Symposium held in London in September 1966. The first conference that aimed to examine destruction in art (not destruction of art) in light of the contemporary cultural setting, the symposium brought together a large group of artists across a range of media who were interested in destruction, as well as a small group of Fluxus artists who were working with destruction and music.

Destruction in Art and Fluxus

In the mid-century, especially during the 1960s, objects were attacked in the name of art. This occurred, for example, by burning, cutting, tearing, slashing, scratching, nailing, dismantling, smashing, breaking, snapping, and shooting. Destruction came to play an important role in the development of new art movements such as Fluxus, Happenings, and performance art, all of which used destruction as a creative process. Fluxus took a central role in this new direction.

Violence directed at musical instruments is a well-documented part of the Fluxus repertoire. This violence ranged from the gentle destruction in Philip Corner's *Quiet Work of Destruction*, which called for the soft and slow dismantling of a piano, to rather more extreme versions in work by George Maciunas. Maciunas' most well-known work took aim at the piano. The piece, *Piano Piece No.13 for Nam June Paik*, also known as *Carpenter's Piano Piece*, asked the performers to nail down all the keys on a piano.[58] The work is an incredibly violent act of destruction. As the nails are driven into the keys the piano starts to fall apart, the keys splinter and the veneer comes away. The sound of nailing, tapping, bashing, and smashing as the keys and piano strings break, as well as the resonances made by the piano that these actions cause, is the content of the work.

Nam June Paik, to whom the piece above is dedicated, is known for his own destruction of pianos and violins. Paik believed that destruction was as important as construction and as such he caused much damage to musical instruments in the late 1950s and early 1960s.[59] In 1962 Paik performed *One for Violin,* in which he raised a violin excruciatingly slowly above his head, before smashing it down onto the table in a flash. The sound of splintering wood and twanging strings would have filled the air as this precious instrument was destroyed. In another work for violin entitled *Violin with String* (1961–1975) Paik pulls a violin behind him, dragging it along the road. Paik himself stated that the dragged violin makes a very beautiful sound.[60]

Paik's more sculptural work *Klavier Integral* (1958–1963) involved the ornamentation and destruction of several pianos. These pianos were decorated with barbed wire, dolls, photographs, toys and eggs, amongst other things. As Ina Bloom observes, "Paik's preparations were more like mutations. The piano, seen as an object of musical culture, was submitted to the violence of transformation as the instrument now reappeared as a sort of matter capable of becoming 'anything.'"[61] In addition to Paik's deliberate modifications, Joseph Beuys destroyed one of the pianos with an axe at the exhibition opening.

The act of destruction in the hands of Fluxus artists serves a number of outcomes. It certainly calls into question the value of the targeted cultural artifacts, as all destructive acts tend to do. It also directs an audience to consider the new, postdestructive act. Does the very act of destruction create something new? Where does this new object lie? Is the new object fleeting—for example, as the unique and singular sounds of the splintering violin? Or does the act of destruction actually create a new object that extends the possibilities beyond those intended for the instrument? I would argue that through the destruction of playback technologies, new instruments and sound possibilities are indeed created. In the example of Cage's *Cartridge Music*, the tone arm wrenched from the deck of the turntable and stuffed with a pipe cleaner makes possible the amplification of the usually inaudible vibrations of the object drawn across a table. Thus destruction is extended beyond the shock of the act and into the arena of productivity and creation. It is with this idea in mind that the Destruction in Art Symposium is to be considered here.

DIAS (Destruction in Art Symposium)

> Destruction art is interdisciplinary and multinational, combining media and subject matter . . . An attitude, a process and a way of proceeding, destruction art is both reactionary and responsive; it is not an aesthetic, nor a method, nor a technique.
>
> —Kristine Stiles[62]

The Destruction in Art Symposium (DIAS) was held in London in September 1966. More than fifty artists from Europe, Japan, and North and South America attended the symposium and related performance events. In addition another fifty artists sent works to be read or performed during the symposium.[63] Kristine Stiles explains that the symposium covered destruction in art over several areas, including:

> destruction of material, destruction caused by nature, destruction of the diachronic, narrative aspect of language, and psychological and physical attacks on the body. The majority of objects and events at DIAS were constructed to be destroyed through various means such as explosives, burning, tearing, and cutting.[64]

During the symposium itself artists gave talks and performances in front of a small audience numbering from twenty to one hundred. DIAS was never meant as a unifying movement but instead drew in artists who were for the most part pursuing their own creative-destructive practices.

Stiles makes clear the idea that this is "destruction in art" and not "destruction of art." This is not destruction of the art artifact; instead, destructive practices are used to create an actual object or an event. Destruction itself becomes the subject matter for many of these works.[65]

Gustav Metzger, the festival's organizer, set out the basis for an "Auto-Destructive Art" in his five manifestos written between 1959 and 1964.[66] As Stiles writes, these manifestos can be read as a "programmatic effort to confront destructive processes in society as well as the dialectic of destruction and creation in art."[67] Metzger, a pacifist, antimaterialist, and political activist, saw future art taking the form of public art aimed at the degeneration of painting or sculpture. These would be civic works that would implode in on themselves through the use of natural forces, traditional art methods, and technological techniques. He also mentions in this first manifesto that the amplified sound of the auto-destructive process could be an element of the work. Metzger's work is highly political in nature, directed toward issues around nuclear weapons, the production of surplus and its links to starvation, and an explicit critique of capitalism.[68] Metzger saw a direct link between a society heading toward mass destruction in the form of nuclear war, and various art movements and artists who used destruction in their artistic process. Metzger acknowledged that earlier art movements, such as Futurism and Dada, had been involved in destruction; but to him it was only in the 1960s that destruction was systematically incorporated into artistic practice, and therefore it was important to bring together artists involved in destructive processes at that time.

Before the symposium George Maciunas wrote a letter to DIAS stating that Fluxus was not about destruction.[69] That said, there were numerous Fluxus related artists either in attendance or who sent work to be read or performed at the symposium, including Giuseppe Chiari, Philip Corner, Robert Filliou, Al Hansen, Milan Knížák, Ben Vautier, and Wolf Vostell. Of most interest here are Philip Corner and Milan Knížák, owing to the nature of their work in music and sound.

Corner wrote *With Violent Takeup No Transfer* for DIAS, to be performed by pianist Ferby Buonanno. The work called for a violent action carried out on near-indestructible objects by tools that would impede any destruction. Much force was to be focused on an action with little effect, for example, a violent attack on a rock with a bare hand. The piece was not performed, however, as Buonanno did not have a piano as had been planned, as it was destroyed earlier by Rafael Ortiz (a performer who often destroyed instruments and live animals in performance).

Two additional works by Corner, *Reconstitutional* and *2nd Finale*, were sent to DIAS. Both involved a gentle destruction.[70] *Reconstitutional* was a duet for a mechanic and a piano tuner. The mechanic was instructed to wreck, with a wide variety of tools, the sounding parts of the piano that were capable of immediate repair such as the strings and pins. The mechanic was to do this subtly in such a way as to impair performance but not totally destroy the instrument. The tuner's instructions were to restore the functioning of a piano as quickly as possible. The tuner was to replace strings and pins and tune the piano during the course of the performance. In the score the mechanic is asked to do his or her part with restraint, so as to allow the tuner to do his or her job. Also when a string is broken the mechanic is not to break any more strings until it is immediately replaced by the tuner. While the tuner is actually tuning the string the mechanic is to go about his or her part quietly. In the end the piano is to be fully restored to working order.[71]

The combination of the destruction of the sounding part of the instrument with the refitting and tuning of the piano would have made for an aurally full performance that uses destruction and repair in a tonally rich manner. Both actions can be seen as constituting extended techniques. The sound of the repair is as much a new and expanded sound for the piano as that of breaking strings and pegs.

Corner had produced other works for the destruction of piano including *Piano Activities,* which was performed at the first Fluxus festival in Wiesbaden in 1962. The work was not originally intended to include the destruction of the instrument, although in this performance the activities were interpreted as such and the piano was taken apart violently with crowbars, saws, and hammers.[72]

Corner was against any wanton destruction, but when informed that the parts had been turned into sculptures he understood that the piano had been positively transformed by its breakdown. Subsequent to the performance of *Piano Activities* Corner wrote *Quiet Work of Destruction,* which called for sensitivity from the performer, who was asked to carry out all destructive actions softly and slowly. These actions were to carried out with patience and attention to small details. Corner even asked that small filings from the instrument be given to the audience.[73]

Stiles has also uncovered previously unpublished information about Milan Knížák's attitude to destruction. Knížák had wanted to attend DIAS but could not obtain a passport to travel, so he sent the symposium documents of actions from the early 1960s and a text he wrote especially for the event. The text describes a type of activity that leads to "the beauty of destruction."[74] One of the works for which he sent documentation to DIAS was *Manifestation of August 2nd*, a Happening performed by the Aktual group, of which he was a member, involving a day of events including the destruction of books, sheets of paper thrown onto the muddy streets and crumpled into dirty balls, and broken objects being taken into a house where a scratched gramophone record played.[75] The group created other similar events that included the destruction of cultural objects such as instruments, paintings, and sculptures.[76]

Knížák had also been realizing *Ceremonies* in which he destroyed objects and ritualistically demonstrated in the streets of Prague. He also produced artist's books that were handmade and characterized by their "torn, burned, and otherwise destroyed and manipulated pages."[77] Interestingly these destroyed books were the means by which much of Knížák's work made it out of Czechoslovakia (see section 4 below, "Milan Knížák's Broken Music," for more detail).

3 Nam June Paik's Broken Tape Loops and Modified Turntables

Nam June Paik is well known as the grandfather of video art. A number of monographs have been published about him, and many of the stories of his early career have been told and retold since at least 1982.[78] Less has been written about his early work, which was primarily focused on music. Within this period of production we find numerous examples of the use of broken and cracked media in the form of old records and modified turntables, tape loops and cut-ups, as well as broken and damaged musical instruments. It was Western music that provided an entry point for Paik into a radical break with his own culture and with traditional Western art music.[79]

Paik was born in Seoul in 1932 and began his artistic career as a composer.[80] After graduating from university, Paik moved to Europe in 1956 and studied at the University of Munich and at the Academy of Music in Freiburg with Wolfgang Fortner. During this time Paik composed music using the collage techniques originally developed by the musique concrète composers in France. He composed a piece based on a Korean poem and included tape-recorded elements such as a baby babbling, water sounds, and snippets of Tchaikovsky.[81] The use of recordings and his own performances of well-known compositions in these tape pieces was repeated a number of times in future compositions, in what was to become an often shocking and confrontational performance style. In 1957 Paik met Karlheinz Stockhausen at the International Summer Courses for New Music in Darmstadt, and the following year he met John Cage.[82] During these years Paik broke away from the traditions informing new music in Europe, specifically the twelve-tone system. While Stockhausen supported his work, Paik was already showing more of an affinity for American experimental music and proto-Fluxus methods than for contemporary approaches to European composition.

Paik's work began to use elements of theater and performance in a Fluxus-like manner a number of years before the official beginning of Fluxus itself. In *Hommage à John Cage* (1959), for example, he used a combination of audio tape and live theatrical performance. The tape collage was made from multiple sources including recordings of Beethoven and Rachmaninoff, a lottery announcement, news reports, toy cars, sound effects, a prepared piano, sine waves, and the screams of many young women.[83] Records were played at the wrong speed, including sounds of sped-up trumpet music, and the amplified pops and crackles from the surface of damaged records could be heard, along with the thump of the needle being dropped on the disc.[84]

Paik used records in many of his performances, as John Harnhardt explains in an essay on Paik's early works: "Paik's actions with records, which included playing and smashing them during performance, were a further extension of his treatment of the recorded medium and the 'playing' of music."[85] Harnhardt is referring to the multiple levels on which Paik performed music, which included the recital of piano music by Paik himself, the use of a phonograph in performance, and prerecorded tape collages played during performance. These approaches directly question the live performance of music and the expected norms of the concert setting.

Paik, in his infamous performance of his composition *Etude for Pianoforte* (1960), played Chopin on one piano and lay down on a second piano. He also cut off John Cage's tie and a piece of his shirt as well as washing Cage's and Tudor's hair with shampoo.[86] The prerecorded sound that played during *Etude for Pianoforte* begins with a screaming voice, perhaps of a baby played at an incorrect speed, along with a slowed-down voice; manipulated moans and sobs play before a section of many roughly cut-up brief sounds. Many of the sounds are overdriven and distorted or have been made from badly recorded or damaged recordings, and the work is extreme in its content and use of disjointed noises.

Paik's first solo exhibition is of key interest because of its use of modified turntables, tape, and televisions. The exhibition entitled "Exposition of Music—Electronic Television," held in Wuppertal in 1963, allowed for the audience to interact with various installations in different rooms, and often it was the interaction with the objects that created the audio content of the works.[87]

Two works in the show are of particular importance in regard to cracked media: *Random Access* (1963) and *Random Access*

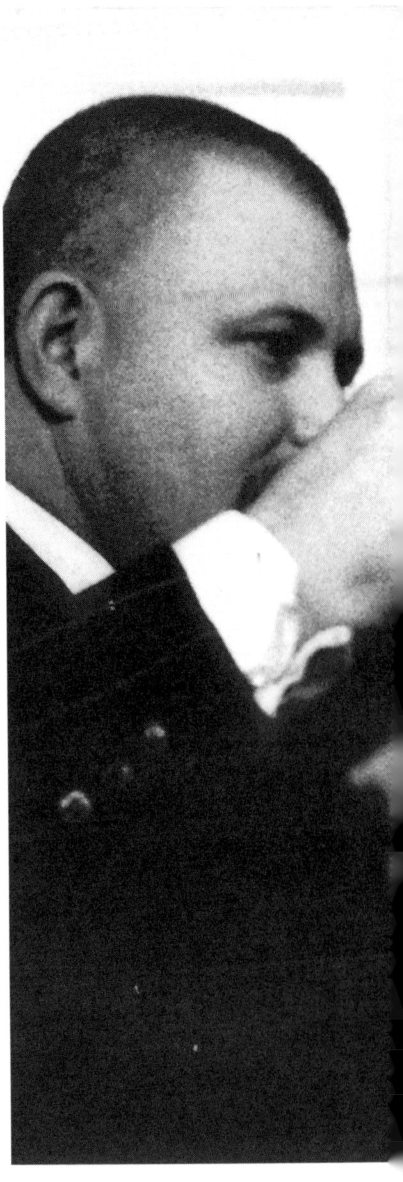

2.2
Wolf Vostell playing Nam June Paik's *Random Access (Schallplatten-Schaschlik).* Exposition of Music—Electronic Television, Galerie Parnass, Wuppertal (Germany), March 11 to 20, 1963. Photograph: © Manfred Leve.

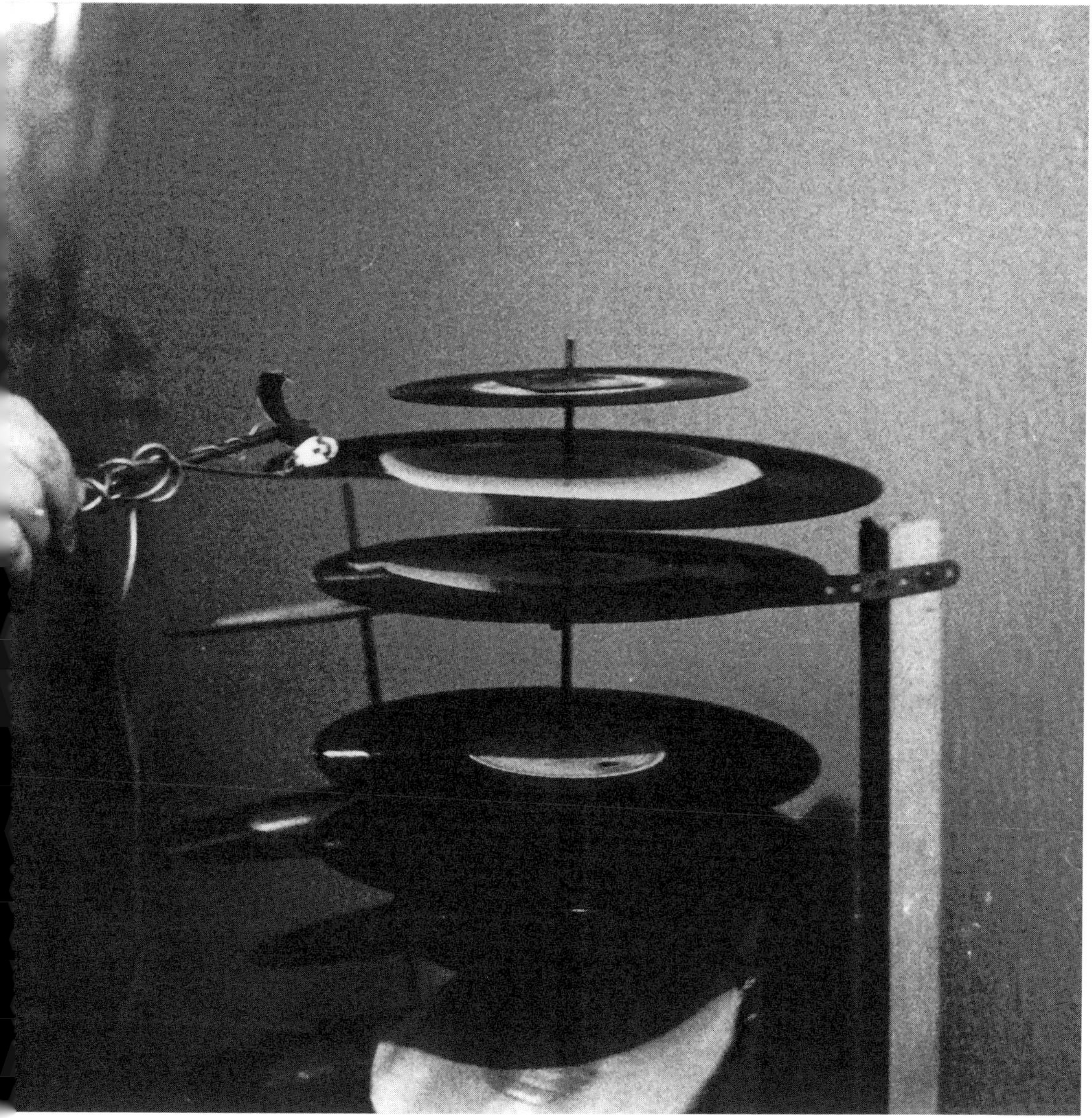

(Schallplatten-Schaschlik) (1963). As the titles suggest, the works allowed "random access" to the typically linear phonographic and tape technologies. The latter allowed the audience to interact with records, a modified record player, and a radio. The phonograph in the work was modified by detaching the tone arm from the player's deck and attaching it to the turntable via an extension. Instead of one rotary plate, the player had a rack of records stacked up via nuts and bolts. Records sat haphazardly at irregular intervals up the spindle's length. The user could then place the needle on any of the records and hence have random access to the various sounds found on them. The work is a wonderfully rickety-looking piece with the various-sized vinyl discs sitting off center, stabilized by a piece of wood attached to the spindle. Paik's work gives users access to over twenty discs at any time, allowing them to jump and skip from record to record, radically cutting up the tracks.[88]

Random Access (Schallplatten-Schaschlik) is directly related to another work in the show simply entitled *Random Access*. In this work Paik cut pieces of magnetic tape and attached the strips to the wall of the gallery. These strips were of various lengths and were placed seemingly at random on the wall, making an abstract image out of magnetic tape. The listener would engage the piece by running the detached tape head across the tape by hand. This is more radical than the work for turntable, as the magnetic tape is by nature linear in its playback method. The listener/user can hear any part of the tape in any order and thus has *random access* to any part of the tape, for any length of time, at any speed. In addition, the piece allows for the demechanization of the playback. That is, instead of the sounds being played in a mechanical, measured fashion, the speed of playback will, because of the nature of the human hand, constantly vary. It would in fact be very difficult to play the tape in a measured way with the hand. The outcome then is a highly variable and complex interactive playback.

These works by Paik extend the playback methods for the turntable and the tape player by radically modifying the technologies through the technique of the crack. His turntables are ruined for normal playback and now require the active participation of the audience to make the technology play a record. By requiring the participation of the audience to activate sound during the exhibition, Paik made clear the active, nontransparent nature of mediation. With random access to the data, the audience creates new versions of the original recordings.

4 Milan Knížák's Broken Music

Milan Knížák was born in Czechoslovakia in 1940 and commenced his career as an artist in 1957. He first created broken and destroyed music in the early 1960s and by 1962–1963 had become part of the Czech art group Aktual, with Sonja Švecová, Jan Trtilek, Jan Mach, Vit Mach, and Zdenka Zizkova.[89] From 1964 to 1966 Knížák created what he called "Ceremonies" and "Demonstrations," Happening-like events that were often performed in the streets.

In 1965 Knížák produced *Killed Books*, a work that involved the destruction of various books. For *Killing the Books* (1965–1970) he instructed the performer to kill the books:

by shooting
by burning
by drowning
by cutting
by gluing
by painting white, or red, or black[90]

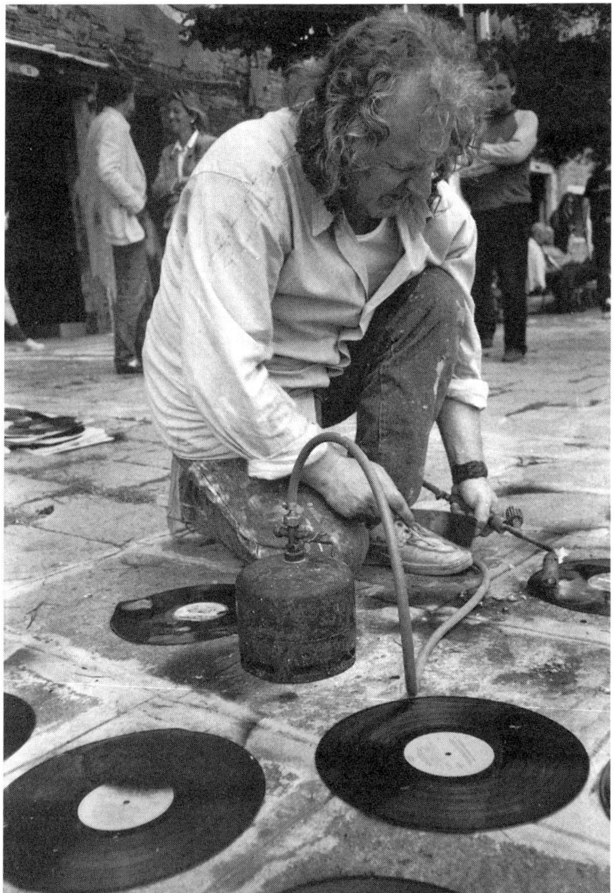

2.3
Milan Knížák working on *Lenin* for the Fluxus pavilion at the Venice Biennale, 1990. Photograph: Marie Knížáková.

Sometime after his release from the army and soon after the initiation of Aktual, he started experimenting with records, mainly, it seems, because of financial difficulties:

> I bought (1963) a record player but there was no money left to buy the records. I had just a few of them. I played them over and over. It was boring. Then I [started] to play them slower and slower and faster and faster. It created funny, interesting, non-boring music. I began to scratch records to make the needle [jump] and so [created] unexpected music or [to make the needle stick] in a scar and [play] the same sequence over and over.[91]

Knížák took his few records and essentially played them to death, in the process creating new music:

> By playing them over (which destroyed the needle and often record players too) an entirely new music was created. Unexpected, nerve-racking, aggressive. Compositions lasting a second or almost infinitely long (as then the needle got stuck in a deep groove and played the same phrase over

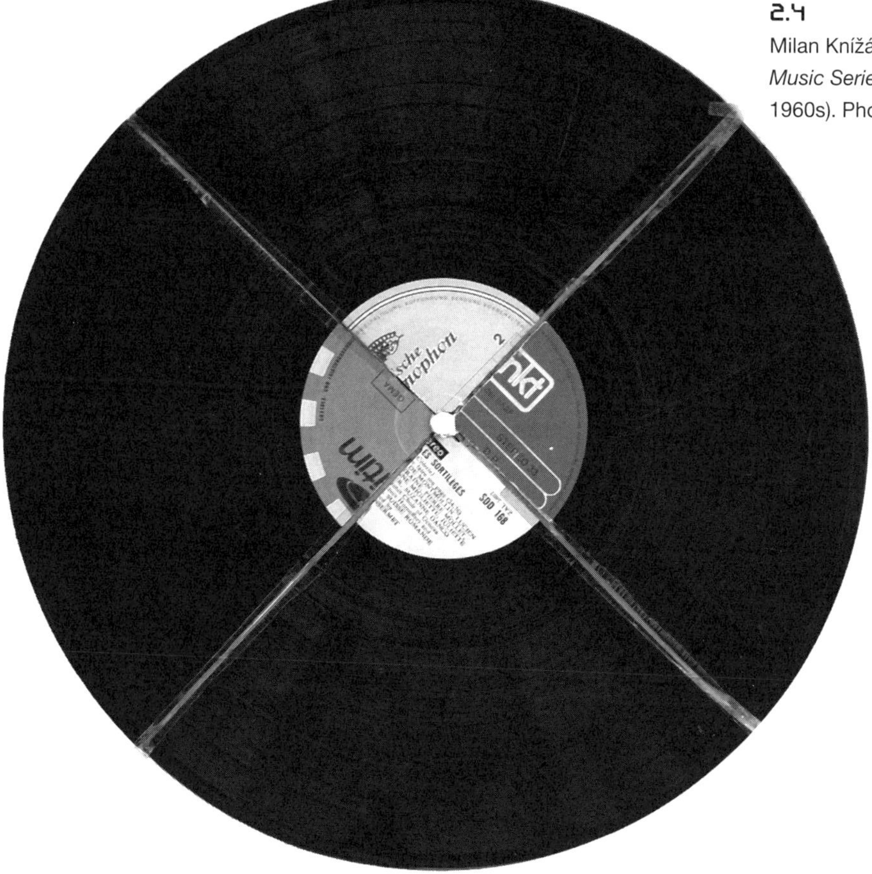

2.4

Milan Knížák, “A Record,” *Broken Music Series*, year unknown (ca. 1960s). Photograph: Milan Knížák.

> and over again). I developed this system further. I began sticking tapes over records, painting over them, burning them, cutting them up and gluing parts of different records back together, etc. to achieve the widest possible variety of sounds. A glued joint created a rhythmic element separating contrasting melodic phrases.[92]

Destroyed Music (1963–1979) includes a work that is made up of four quarters of different vinyl discs.[93] These quarters have been glued and taped back together to form a whole. When the record is played the needle will run through a quarter of a full groove before hitting the join and popping into the next groove, which is a completely different track. Hence one revolution will include elements from all four records. The glue creates a new rhythmic quality to the music; as the needle hits the join it pops and clicks, in this case four times per revolution. *Composition n. 1,* originally released on an album entitled *Broken Music* in 1979, is an example of exactly this.[94] The sound of the breaks forms a regular percussive element in the composition, far stronger than any scratch. These creaks and bangs are heard at regular intervals and change in duration and timbre as the record is spun faster or slower. Between the "beats," the sounds originally on the records play for only a few moments: piano music, violin, and percussion. The overall sound is an ever-changing arrangement of music played as the needle scratches, bumps, and bounces toward the center of the disc. The audio is at once confusing and compelling, as unexpected sounds plays at strange speeds and in disjunction to what is expected.

Another record piece, *Untitled* (1985), takes the form of a cross with slices removed from the originally round disc. In this case only the inner tracks can be played as the outer edges have been removed. In other works Knížák actually glues, at adjacent angles, pieces of other records to the base disc. These records have been violently tampered with, rendering them completely unplayable as shards of vinyl jut out of the base disc. These works are no longer strictly musical, instead seemingly becoming artefacts more akin to sculpture than music. Knížák recalls:

> At the beginning I saw the records just as a musical work and I made only records which were possible to play ([in] some way). In the '70s Giancarlo Politi—the owner of magazine *Flash Art*—took some of my records, sold them to Gino di Maggio who put them on the wall. This was

> kind of a push for me and therefore I have begun to think of records also as sculptures. But I never wanted to give up the musical functions. I pasted or hung on them some small music instruments creating pieces which contain two kinds of possible music: A) recorded, B) live. Records which have pasted wooden or plastic rings, alu-folies, etc. are for "traditional" playing, i.e. the space of a ring creates a small composition, [and the] needle moves only within the space of the ring.[95]

In addition to vinyl fragments Knížák began attaching other items to the surface of the discs. In an untitled work from 1985 he included seven safety pins that were pushed into the surface of the disc. Another disc has a guitar-like bridge and strings attached, and he has also included toy musical instruments such as plastic horns and whistles (*Untitled*, 1963–1985).[96]

Toward the end of the 1980s, Knížák began to apply paint to the surface of his vinyl discs. These discs were purely visual, as the original grooves were covered and filled with paint. They do, however, contain, according to Knížák, "potential music hidden under the paint."[97] This inherent music was to become an element of Christian

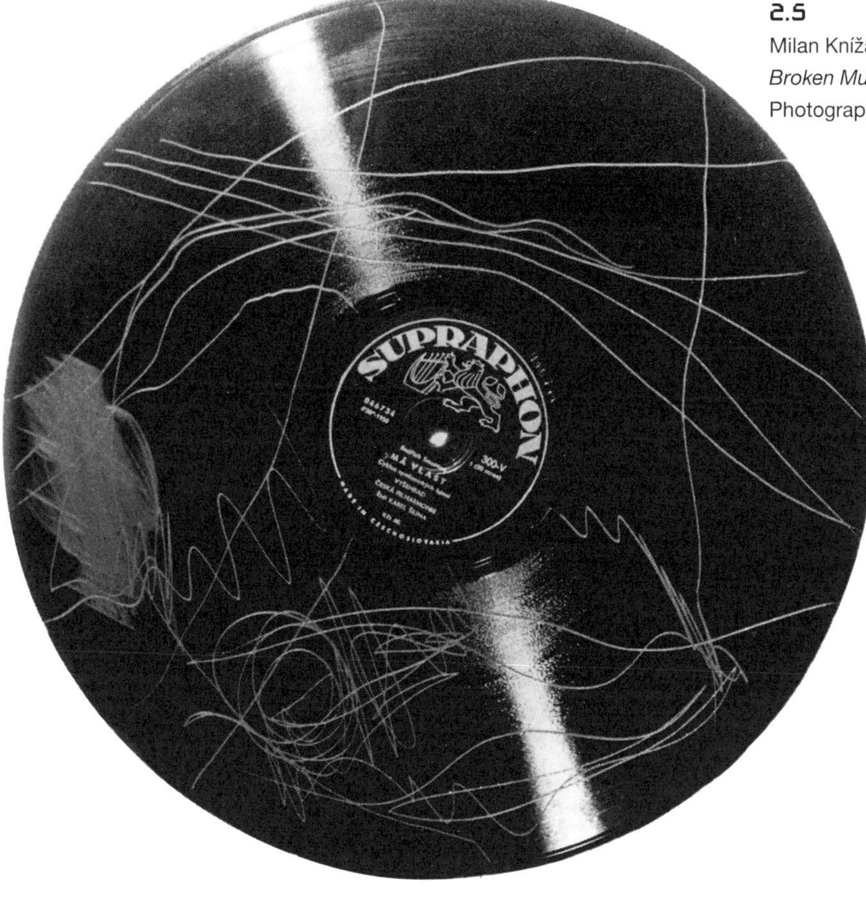

2.5
Milan Knížák, “Scratched Record,” *Broken Music Series*, 1965.
Photograph: Milan Knížák.

Marclay's work, as will be discussed in depth in the following section on Marclay.

In 1988, Knížák took a disc with glued-on record fragments and burned it, melting the disc into a completely unplayable form. When asked about the inherent violence in such actions, Knížák replied:

> I don't understand my working with the records as a violence—it is just a different way of using them. Creating new, surprising, sometimes a bit aggressive, but sometimes also some kind of meditative sounds.[98]

Knížák, in his use of broken music, has taken the record and extended it well beyond the original intentions for the technology, and in the process he has transformed the form of the disc as well as the data stored on it. The music and objects he has created produce new forms of sound, but they still manage to retain at least a memory of the original recording.

Fluxus and Destruction

Fluxus artists such as Paik and Knížák worked with the destruction of phonographs and records, but these Fluxus works are the cause of a major anomaly in the development of cracked media. Fluxus artists attacked instruments with delight, and phonographic technologies were no exception. Until Fluxus the turntable had been experimented with only through the safe manipulation of records and the phonograph itself. These practices contradict the pattern outlined earlier, whereby the modification of recording media through destruction occurred in line with the technology's declining value as a commodity. Why was the turntable, then the major technology for the playback of recordings and a valued home entertainment technology, cracked and broken in this destructive and irreversible manner?

The reason can be found in Fluxus' radical disregard for culturally valued objects. If, in the name of music, a performer can destroy a grand piano, how much easier is it to destroy a phonograph or a record? The scale of destruction is slight compared to smashing a violin or hacking apart a piano. It was not for a decade after Knížák's vinyl experimentation that the phonograph began to be taken up again as an instrument. The next section will look at the vinyl-destroying work of Christian Marclay and the beginning of a fully developed area of practice around the use of the turntable.

5 Christian Marclay and the Death of Vinyl

> I destroy, I scratch, I act against the fragility of the record in order to free the music from its captivity.
>
> —Christian Marclay[99]

Throughout the 1980s and 1990s, Christian Marclay wreaked havoc on vinyl records, abusing them and subsequently redefining them as art objects. He has scratched, sanded, cut, drilled, snapped, melted, smashed, thrown, and walked on these fragile and once fetishized objects. Marclay began using the turntable in 1979, taking up the record and phonograph as an instrument only a few years before the technology began its descent into obsolescence. Compact disc technology was released in late 1982, pushing the market for recordings into the digital realm. By the 1990s the vinyl disc was utilized only by collectors and DJs, the latter keeping production of the format alive, although only in extremely limited quantities. Marclay has continued to use vinyl, the obsolescence of the format allowing him to create his music cheaply and easily owing to the thrift-shop accessibility of the medium. Through his audio, performance, and fine art practice he has extended and expanded the use of the turntable for experimental purposes. In the process he initiated a style of music grounded in experimental "turntablism."

This section aims to address the dearth of critical attention given to Marclay's nonexhibition based practice, focusing almost entirely on his music and performance (although this cannot be completely exclusive as there is much crossover between aural and visual practices in his work). After briefly looking at the varied influences on Marclay's practice, the section will look at his accidental sound-inducing techniques and his approaches to sound creation, including their differences and commonalities with hip-hop turntablism. The second part, "Vinyl: The Record as Artifact," focuses on issues around the objecthood of vinyl and turntable technology in Marclay's practice.

Christian Marclay's Sound Practices

Christian Marclay studied visual arts in Boston at the Massachusetts College of Art before moving to New York in 1980. He had lived in New York City in 1977–78 while an exchange student, and during this time he was exposed to performance art, punk, and no wave, which were to become important influences on his practice. Marclay acknowledges the "Fluxus attitude as an influence," especially in what he reads as their ridicule of and play on the formal presentation of music in the concert setting.[100] This attitude can be seen in concerts performed by the Fluxorchestra between 1964 and 1965. Many of these works had humorous elements to them: the performers dressed in the formal attire required for a concert of "high" music while the conductor took on the often ludicrous position of leading the musicians and performers through slapstick-like routines. Balls of paper and paper darts were thrown around the theater, musicians fell off their chairs, while the conductor seems to prepare for the performance by tidying himself up only to end when he should be ready to begin.[101]

In an interview with Vincent Katz, Marclay cites John Cage's acceptance of noise and the accident as an influence on his own work.[102] Specifically, he refers to the idea of putting sounds together in a random (indeterminate) manner. These were extremely important developments for Marclay's music, which mixes and matches unrelated sounds, noises, and breakages in performance. Marclay asserts that the futurists' noise and Marcel Duchamp's accident were influential, as well as musique concrète's manipulation of found and recorded sound.[103]

When Marclay was a teenager he saw a work by Oliver Mosset in which the artist posed in a black leather biker's outfit in front of a painting. Marclay remembers that the "juxtaposition between art and biker culture surprised me at the time. It seemed that youth culture and art did not belong together, but since then I've learned to love the blurring of these differences."[104] It was this very combination that Marclay brought to his own practice as he moved between musical styles and art practices. In an interview with Jonathan Seliger he states that along with "everybody starting a band," punk and no wave had a significant influence on the New York art scene at the time. Many of the bands came directly out of art school, leading to a productive cross-fertilization between music and art.

Marclay connected the art scene with the New York downtown improvised music scene. In the 1980s he appeared on many studio albums with groupings associated with the musician John Zorn in his "game pieces." The list of performers he has played with is vast, especially considering that Marclay is not a trained musician but rather a visual artist.[105]

Marclay is also an exhibiting artist, a practice that both informs and is informed by his musical performance. His sculptural sound art is full of references to music and the music industry, carrying on a dialogue very similar to that of his music practice. His art practice includes the use of records as sculptural pieces. For example, in *Footsteps* (1989) he placed 3,500 records on the floor of a gallery for patrons to walk on; he made collages of record fragments, originally to be listened to, for the works *Recycled Records* (1980–86); and he drew out implications of image, gender, and fashion in the music industry in his *Body Mix* series (1991–92). With Marclay we are never far from sound.

There is a large body of writing about Marclay: two books[106] and numerous interviews, articles and reviews. The vast majority of this writing is about Marclay as an exhibiting sound artist, with very little focused on his audio and live performance. His work has been very well known in the international art scene since the late 1980s as evidenced through a number of high-profile shows, including the 1995 and 1999 Venice Biennales, the Solomon R. Guggenheim Museum in New York, the Whitney Museum of American Art in New York, the Museum of Modern Art in San Francisco, and the Museum of Contemporary Art in Los Angeles. Whereas his art practice is clearly part of the institution of contemporary art, his music is much more difficult to categorize, and because of its experimental nature it has not been recognized to the same extent as his art practice in either popular or classical musical circles.

In many ways Marclay is a quintessential postmodernist of the 1980s. His use of samples, found sounds, debris, and old technologies is a bricolage of incongruent cultural elements, which, in his hands, become ripe for exploitation. Marclay openly moves between the worlds of the high and low arts, as comfortable improvising with musicians in a cafe as exhibiting in art museums. At one turn he is at an art opening in Europe walking on his own vinyl records; at the next he joins cult art rockers Sonic Youth on their *Goodbye 20th Century* release of old favorites of experimental music.[107] At the center of this practice is the abused vinyl record and the record as object, and it is Marclay's handling of the record disc that has made for his continuing success as a practitioner of sound.

Technique

> Marclay is a born showman, slapping one album onto the platter, sailing another over the edge of the stage to shatter at the audience's feet.
>
> —Mark Dery[108]

Marclay began performing with turntables in 1979. At the time he was singing in a punk duo called The Bachelors, Even, formed in 1979 with Kurt Henry. The band's name is something of an art school in-joke, referring to the painting by Marcel Duchamp, *The Bride*

Stripped Bare by Her Bachelors, Even (The Large Glass) (1915–1923). It places Marclay firmly as an art school student while nodding to Duchamp as an important influence, especially on Marclay's fine art practice.

The Bachelors, Even used guitar, vocals, and tapes, which Marclay made from sounds produced by playing damaged records. He later started to use records instead of tape: "I used them [turntables] like an instrument, and could adapt my playing to a live situation, it allowed for a lot more freedom and spontaneity than tapes."[109] Since this time the turntable has been the mainstay of Marclay's musical practice.

At this time he also began experimenting with the "Phonodrum," a rhythm machine consisting of "a short piece of guitar string taped to the tone arm as an extension of the needle, the string drags across a record or a wooden disc riddled with nails and bounces around. The vibrations are picked up by the needle and are highly amplified."[110] The amplified guitar-string needle made mid-range twanging sounds as it bounced rhythmically against the obstacles attached to the record's surface. In future developments of his technique Marclay focused primarily on prepared records and extended techniques for playing them, rather than preparing the turntable itself as in the Phonodrum.[111]

A few years prior to Marclay's first experiments with records, Laurie Anderson, an acknowledged influence on Marclay, made a series of pieces using altered violins fused with phonograph technology. One such work utilizes the "Viophonograph," which was made in 1975 by Anderson and Bob Bielecki. A battery-powered turntable was fitted into the body of a violin. A vinyl 7-inch disc replaced the strings and sat where the bridge of the violin would normally be. The bow itself had a needle attached and the needle was held onto the spinning disc much like the conventional turntable action. The needle could be placed at any point on the disc. Anderson made a disc with one note on each groove of a 45-rpm single; thus she could produce notes by placing the needle at various points on the disc while mimicking the action of a conventional violin performance style.[112] The Viophonograph was one of a number of violins made for Anderson, including a Digital Violin (1985), Neon Violin and Neon Bow (1982, 1985), and the Tape Bow Violin (1977), which used a section of tape instead of horsehair on the bow. In the later work the bow was dragged across a tape head attached

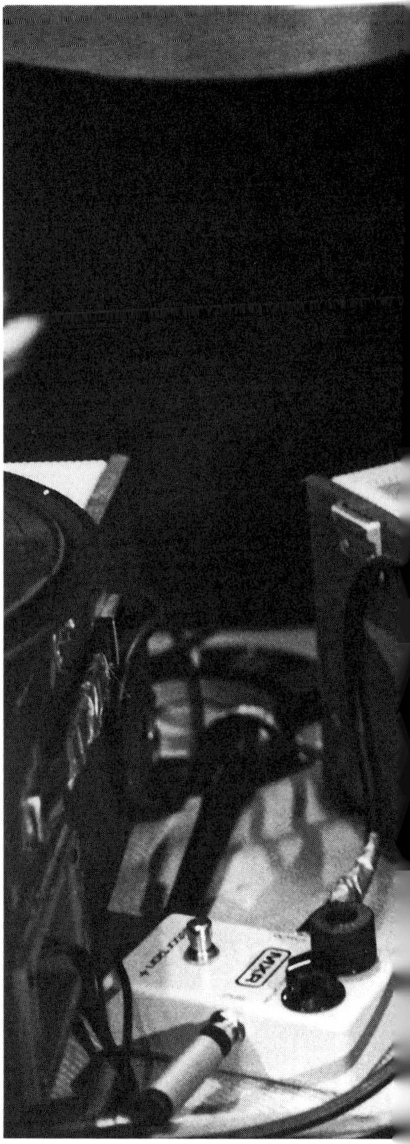

2.6

Christian Marclay with Phonodrum, 1980. Photograph: Kei Orihara.

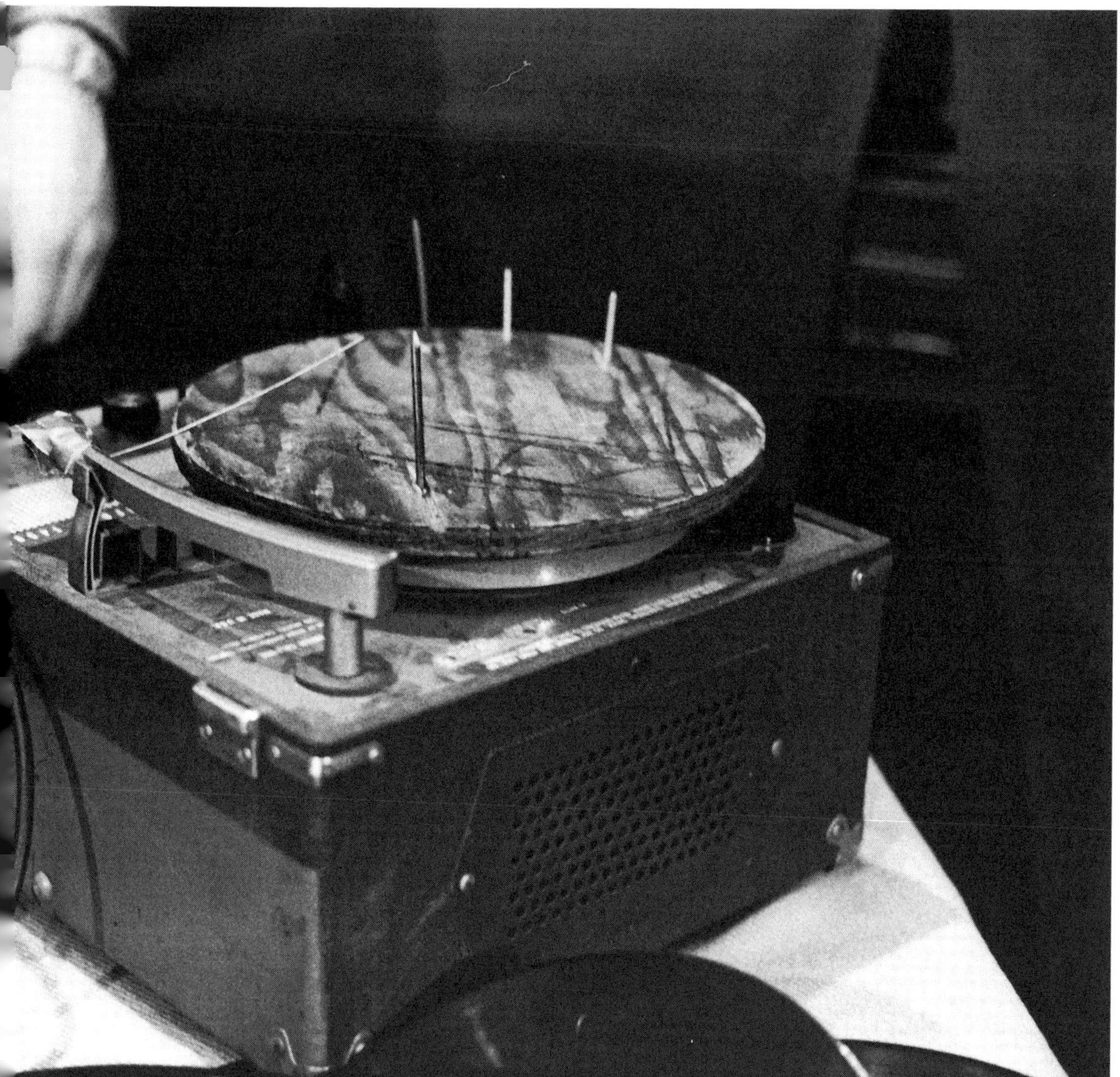

to the violin body. Both the Viophonograph and the Tape Bow Violin resonate with Nam June Paik's *Random Access* works, in that they remove the mechanisms of the original players and allow the user to have free access to the data stored on the media.

Marclay's audio output is made up of "slabs" of sound. The audio is taken from vastly different sources that are then "cut up" and juxtaposed. The overall sound is a combination of what are usually heard as completely disparate musics: cuts from discrepant recordings, Tin Pan Alley recordings, classical music, waltzes, severely cut up drum solos, and disco records. In the early 1980s Marclay's technique was anything but familiar. He tore through records in live performance, producing loud abrasive scratches, the rasp of a needle drawn across the vinyl, spinning record discs back and forward. In addition to using the techniques of the DJ, the mix and scratch, Marclay introduced severe vinyl damage and destruction to his repertoire. His work from this period still sounds extreme in its use of noise, although recognizably of the period. This is partially due to the overt postmodern nature of Marclay's practice and the use of 1980s music in the actual pieces.

Marclay uses a Califone turntable, the model that was used in schools across America for record playback, built to withstand the abuse of young students. These devices were not designed for DJ-ing but they proved to be useful for Marclay's purposes. These turntables have four speeds, and Marclay has at times modified them by adding an extra tone arm and on-off switches. The turntables themselves are large and thick, with bulky metal tone arms. He has said that he likes "the old-fashioned turntable: because it is so dumb. You can hit it, you can do all these things, and it will never stop playing."[113]

Marclay prepares his records, modifying them to produce additional audio through their malfunction. He scratches, cuts, and buffs his records to create an audio texture thick with rasping, ticks, and pops. He has stuck scotch tape to the records, in the form of both thin strips attached through the center of the disc and squares positioned on the records seemingly at random. Like Milan Knížák, Marclay literally cut up the records, slicing them into quarters, mixing up the pieces and re-forming them. The join in these *Recycled Records* cannot be perfect, causing a loud click as the needle hits the glued break. As the grooves do not quite line up, the reaction of the needle to the cut is extremely unpredictable. Marclay comments, "As they were played, the needle would sample from different fragments of records. The music was loud and gritty."[114] In addition to using the *Recycled Records* in performance, Marclay turned them into one-off sound objects.

Milan Knížák did play his reformed records, but this was carried out simply by allowing the needle to play a single work at time. This is very different from the manner in which Marclay performs his discs, utilizing multiple turntables and multiple record works to create a performance event. Another difference between the two is in the playful nature and lightness of the sounds used by Marclay. Knížák's audio is thick with serious sounding classical music while Marclay's sounds come across as humorous and light-hearted. *One Thousand Cycles* (1981), for example, has a fast rhythm and includes fragments of voice and slices of melody. The rhythm created by the cuts themselves is upbeat and there is a certain levity in these pieces.

The majority of Marclay's live performance techniques fall into the category of prepared instruments, in this case prepared records, directly linking his practice to the twentieth-century preoccupation with prepared and extended techniques. Gary Butler discusses two types of preparations that are of direct relevance to this discussion, "indeterminate" and "destructive."[115] The use of preparations can yield a factor of unpredictability in the musical performance: at one extreme the performer deliberately creates a situation where he or

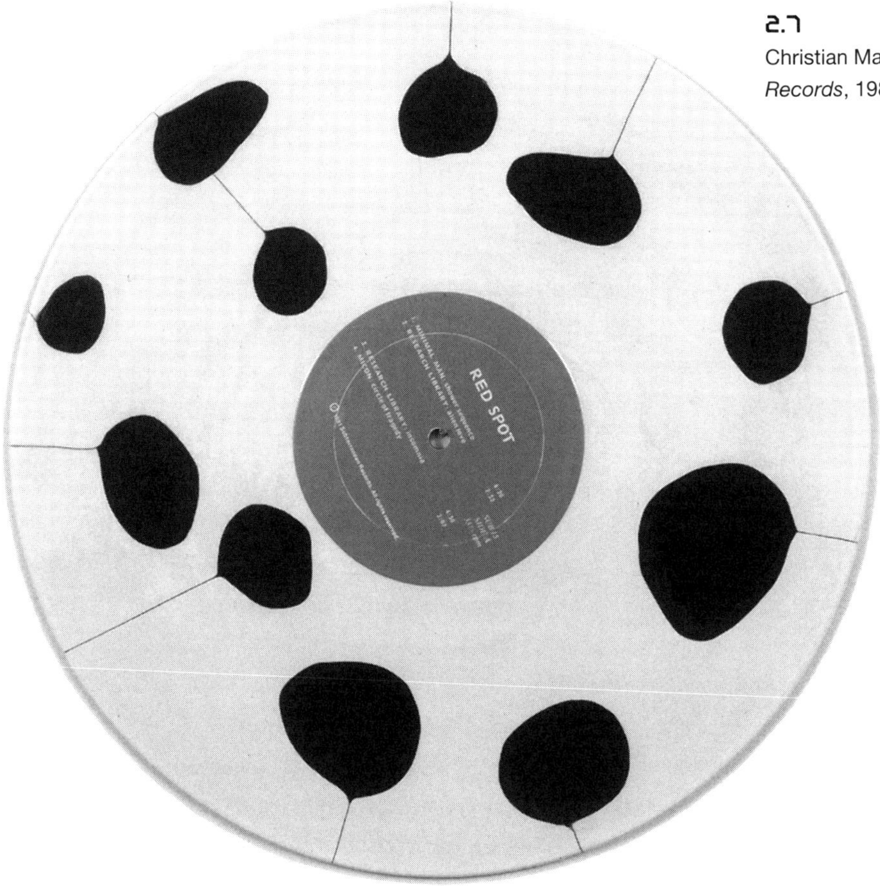

2.7
Christian Marclay, *Recycled Records*, 1986.

she can have no idea of what will come about, while at the other there is a more managed and almost predictable outcome. The accident becomes a positive aesthetic, something that is sought after. Cage, in his early use of the prepared piano, formulated a highly controlled use of the objects and even attached to the score the actual objects to be used in the preparation of the piano. Later he realized that the preparations sounded different from piano to piano and that in fact the very use of the preparations themselves needed to be seen as indeterminate. This idea was more in keeping with his later ideas about the indeterminate in composition.[116] The prepared piano is a part of the trajectory of extended sound for the piano, allowing new sounds to be drawn from the instrument. In the same way the prepared record does just this; it generates sounds not originally intended by the record producer.

The use of destruction in preparation and the permanence of a preparation is also relevant here; once Marclay has prepared a record, there is no going back. It cannot be returned to its original form and is often finally thrown out into the audience as shards of smashed vinyl. During the performance of prepared records the needle glides around the spiraling grooves, hitting the various blockages, slipping, skidding, and sliding across the pieces of tape.

The out-of-control needle might slide across the grooves much like a stylus that has collected a pile of lint from the surface, careening into the runout groove and producing a tearing sound. Perhaps it hits the tape and makes a rhythmic tick as it passes over. Not only is the incident outside the original intention of the recording artist, but the outcome cannot clearly be predicted by Marclay. The semi-random event opens possibilities for unintentional performance and semi-intentional improvisation. Marclay states:

> When something goes wrong, like when the needle skips, something unpredictable happens, that wasn't the intention of the recording artist. In that incident, something new and exciting happens. For me, it has creative potential.[117]

Thus the outcome of the experiment is unclear. In this case the performer is happy for these accidents to occur; in fact, the tape is added to the records to make certain that accidents do happen. This is a fundamental difference between Marclay and the DJ who mixes for the dance floor. If the needle were to skip, jump, or slide across the vinyl during a set the DJ would be shocked and embarrassed, correcting the problem as soon as possible so as to allow the crowd to get back in the groove. Marclay looks for and sets up such errors and unexpected occurrences, finding creative potential in them.

Marclay and Turntablism

One cannot look at the use of contemporary turntables without considering hip-hop. Throughout its history, hip-hop has invented and developed various techniques and approaches for the use of the turntable as an instrument. The virtuosic high point is reached in the abstract display of technique known as turntablism. As Ulf Poschardt observes in *DJ Culture*, the turntablist mines the entire spectrum of recorded music in a hunt for the "breakbeat" or the perfect sound bite to sample.[118] The DJ has available to him or her all recorded sounds and in the process of spinning records will mix and scratch prerecorded sounds to create new and original music.

Turntable virtuosity developed out of the simple drive for the perfect breakbeat and its use on the dance floor. Bill Brewster and Frank Broughton describe the effect of the virtuosic turntable manipulator:

> Hands whipping from one record to the other, stopping lightning fast on the crossfader in between, shoulders

> dipping slightly in time to the beat . . . fingers moving in millimetre-precise formation, each flick or slide or rub controlled to a hair's breath, and from the speakers a pounding beat with a barrage of scribbles and scratches.[119]

In the 1990s turntablists such as DJ Q-Bert, Mixmaster Mike, and the Invisibl Scratch Picklz became known for awe-inspiring feats and moves. The scene revolved around the invention of new moves and tricks, much like the skateboarder showing off a new trick on the half pipe.

In an interview with Jason Gross, Marclay discusses the relationship between his own practice and hip-hop. For him the similarity is the use of "records as instruments to create new music out of old."[120] He points out that whereas hip-hop is interested in dance music, a practice he is not engaged in, it was hip-hop that really made DJ-ing an "accepted craft." Like the trick-based turntablist Marclay is not interested in getting people to dance, and he has very different influences from those of hip-hop, deriving from experimental practice in both music and the visual arts. Marclay hears the use of scratching in hip-hop as a way of accenting the groove. These techniques developed out of an interest in heightening the experience of the dancing crowd. As Marclay remarks, "Hip-hop DJs want to get people to dance and keep them dancing, somewhere between sociology and crowd control."[121] There is no interest for Marclay in a steady beat; in fact in performance he goes to great lengths to make sure this is not possible. His scratching is not timed to accent the groove and he prepares his records so as to make certainty an impossibility. Thus despite their shared use of the turntable, the two musical practices are related to two different and divergent cultures and the two scenes have very little to do with each other.

There are also similarities, however, between the destructive techniques of Marclay and the experimental edge of techno. The actual cutting of new grooves into a vinyl disc is not uncommon in dance music culture, and other producers have used slashed and cut records as a DJ tool to add rhythm and texture to their sets. Although much later than Marclay's original vinyl destruction, a number of minimal techno producers have used techniques similar to Marclay's to make stripped-down machinic techno. German artist Thomas Brinkmann, for example, cuts grooves into records and plays a modified turntable with two tone arms. He describes the process:

> I would take a regular record, the last loop [the run-out groove], and I would cut formations with a sharp knife. . . . You make a cross on the label, and at the end of each point from the cross you make a deep scratch. When you play the record at 33 rpm, you have a beat like a bass drum at 133 bpm. "Boom, boom, boom . . ." . . . And if you put, between those four points, four little scratches, you have exactly the grounds for a techno track. "Boom, tic, boom, tic, boom, tic. . . ."[122]

Although Marclay does not perform with the intention of holding a regular beat, when offstage his process is similar to that of the breakbeat-obsessed hip-hop DJ who scours the second-hand record store. These DJs hunt and collect by looking for double copies of new and unheard "breaks" to add to their hidden archive. The DJ as archivist is at no point more extreme than here, the fruits of their labor hidden from the eyes of the public as these sought-after grooves are made into beat sequences for new DJ sets.[123]

Similarly, Marclay produced an extreme collection for an installation at ArtPace in San Antonio, Texas, acquiring what he called “an absurd archive” of over 1,000 LPs of Christmas music. For the opening and during the exhibition, DJs were let loose on the collection, mixing the Christmas cheer for gallery goers (Marclay himself mixed at the opening).[124] The records were placed in cardboard boxes and cataloged around various themes: compilation albums, alphabetical by artist, international Christmas albums, rap Christmas albums, and different versions of *The Little Drummer Boy*. Here we have the Christmas album, which epitomizes “light” music, up against the DJ mix and scratch, which is often abrasive and harsh.[125]

Vinyl: The Record as an Artifact

Christian Marclay’s performance practice is centered around the vinyl disc and its objecthood. By using and abusing the record Marclay draws attention to the process of sound creation. Thus the “meaning” of his work has greatly changed since he began using records and turntables as instruments in 1979, as the meaning of the record itself has changed. For most of the 1980s, the record was the main method for the reproduction of recordings and personal playback. The vinyl disc was inscribed with a quality of preciousness and reverence. The object was handled with a fetishistic care, with users continually waging a losing battle to preserve their precious discs. The record itself wears out further with each successive playing as the needle, running along the grooves, wears down the information held within. In addition, dust and small scratches produce pops, pocks, and ticks while larger scratches produce rhythmic ticks and snaps, "ruining" the listening experience. Marclay remembers when he first noticed the faults in the playback of vinyl discs:

> I realized that when I listened to a record, there were all these unwanted sounds, clicks and pops, because of the deterioration of the record, the surface noise, scratches. Instead of rejecting these residual sounds, I've tried to use them, bringing them to the foreground to make people aware they are listening to a recording and not live music.[126]

Marclay exposes these faults in his performance, treating his vinyl with little concern. The music loving vinyl collector must have been shocked to see Marclay handle his discs in such a way. However, our contemporary approach to the medium is vastly different, as the phonograph has become, in a very real sense, obsolete. In the twenty-first century many people have in fact never handled a record let alone used a turntable. Marclay on returning to America was surprised by the excess and waste of the culture.[127] He saw "records on the street, in the gutter. I would see thousands of records in thrift shops that nobody wanted, that nobody cared about. It was in some ways that cultural change that allowed me to see a different attitude towards records, and I pushed in that direction, considering them as just a cheap commodity to be used and abused."[128]

Marclay's use of the breakdown in the technology is a demonstration of the impermanence of the medium. He also exposes the myth of perfect and transparent sound reproduction, in a manner similar to that outlined earlier. When listening to the phonograph it was necessary for users to tune out the extraneous noises of the technology. Marclay, however, hears the pops and ticks, especially on old and discarded record discs, as material for his music and questions the desire for perfect sound reproduction. This is no clearer than in his record-based installation practice.

In the 1980s Marclay released a number of albums that were directly related to his installation practice. One example is *Record Without a Cover* (1985), which was sold without a sleeve, and the buyer was instructed not to place it in one:

With *Record Without a Cover* you can't ignore the medium. You can't ignore that you are listening to a recording. There is confusion between what is intentionally recorded and what is damage to the surface of the disc. There's a push-and-pull between the reality and the illusion. You have to stay alert.[129]

A similar example is *Footsteps* (1989). For this piece, he covered the floor of the Shedhalle in Zurich with 3,500 records, which were subsequently sold; the marks and scuffs created by gallery visitors walking over the discs became part of the record's audio. The buyer of the *Footsteps* release thus holds a new object that has been severely damaged, the record's surface having been scratched and scraped by the gallery audience. The pops and ticks sound out over the original recording, making the listener, as Marclay stated, fully aware of the medium on which the recording has been made; in fact, the recording is in a significant sense of footsteps and the damage caused by the steps of the gallery audience. They are multiples, a type of print. They are artworks in which each individual record is infused with elements of indeterminacy: Marclay knows that the

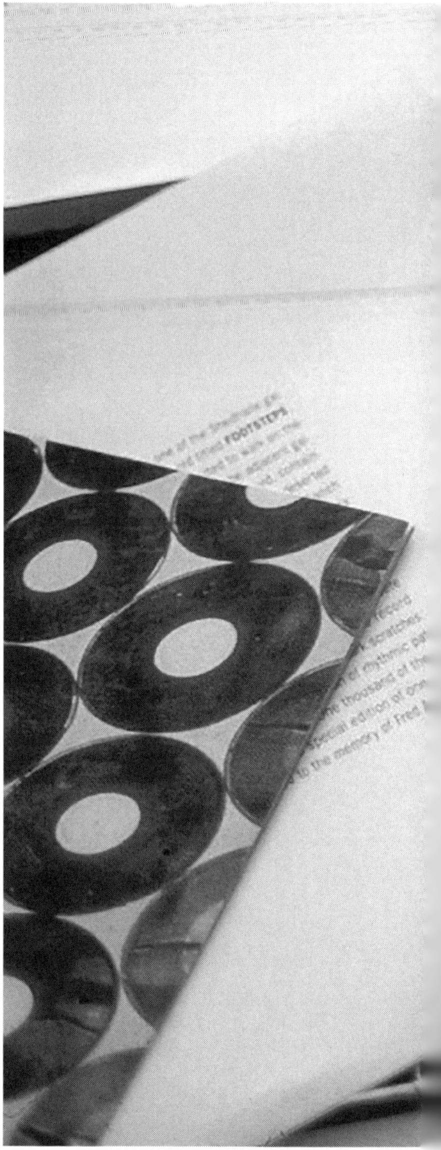

2.8

Christian Marclay, *Footsteps*, 1989.

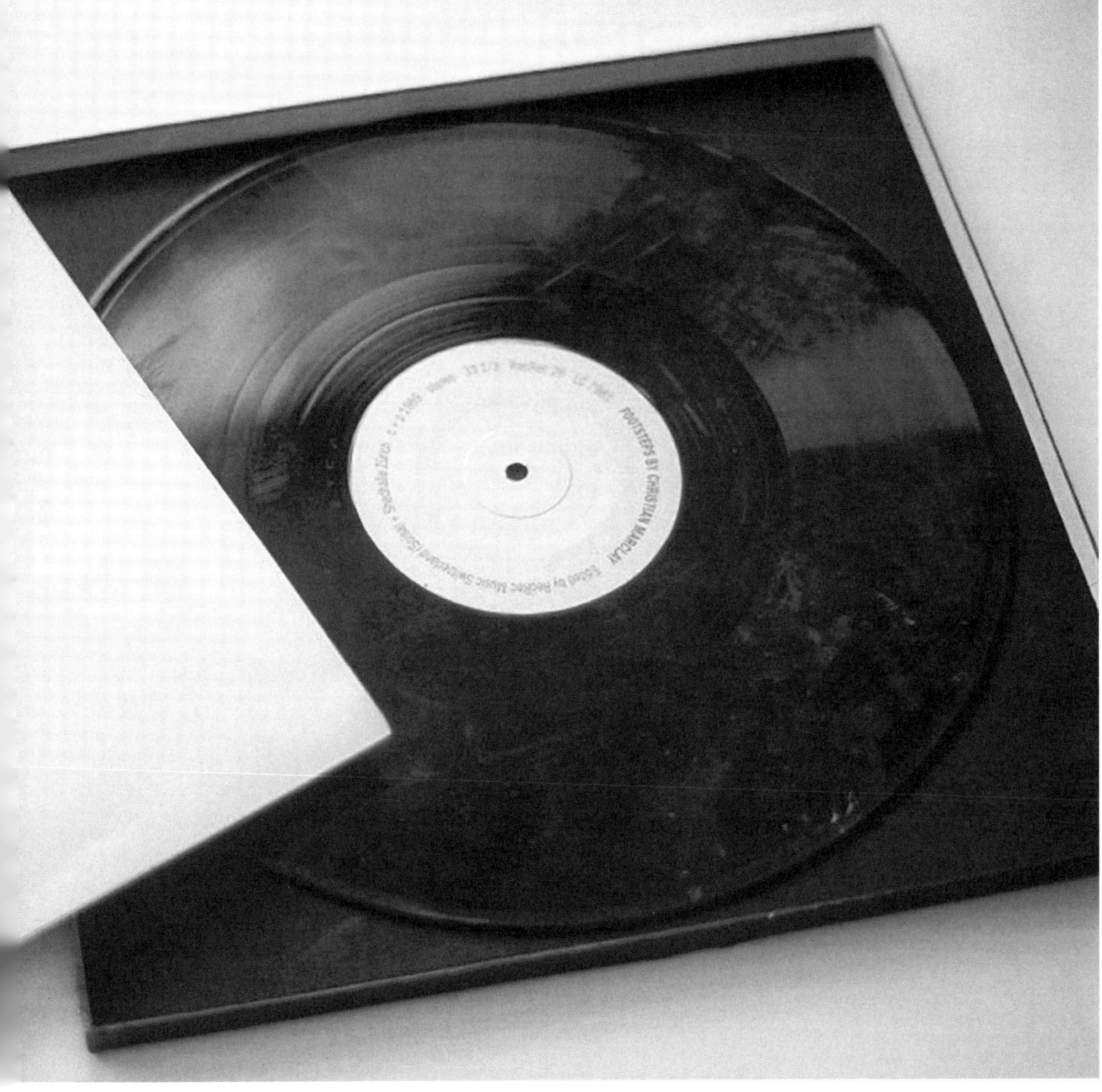
FOOTSTEPS BY CHRISTIAN MARCLAY

gallery patrons will scuff the record, but each individual disc will be differently damaged.

The opposite of these releases of severely mistreated vinyl is an untitled limited edition from 1987.[130] This release is a record without grooves, the vinyl sold in its own suede slipcase complete with drawstring. This luxury collectible is completely useless in that it is without audio, lacking the necessary grooves to guide the needle safely around the disc. Another example of the unusable record is *Secret* from 1988, which consisted of a metal 7-inch disc with a padlock through the center hole and no key to unlock it.

These examples of Marclay's early output of records must be read in the context of the 1980s, when a mark, scratch, or pock to vinyl was something to be avoided at all costs and needles were kept new so as to not run down the grooves. Marclay's approach, then, is twofold: he is commenting on the fetishized vinyl object (recall Eisenberg's position, recounted earlier, that his money hears the record he's just purchased) while at the same time drawing attention to the aesthetic value of the sound, or noise as it was normally heard, that was formed beyond and after the original recording and pressing process.

Although Marclay's practice is part of a conversation about the limitations of vinyl and the phonograph as a technology, as much as it is about questioning the fetishistic desire for the commodity, he is not enthusiastic about digital technology:

> When I use vinyl records, I'm going against technological progress. Vinyl is a kind of hands-on technology, and it's imperfect, it's fragile. It's still mechanical as opposed to a digital technology. I'm talking about the time before digital technology was invented. Then the technology suddenly changed, and you had this new technology that allows you to do the same thing in a very sophisticated way. But in performance, CDs have their limitations. The digital technology doesn't allow you to do live mixes, directly in front of an audience, and create something on the spot, making mistakes, correcting yourself, going through the whole process of creating something with its failures. It's when the technology fails that something interesting happens.[131]

The position of the artist who works with the limitations of the technology is clear here. Marclay is not simply jumping to the newest and best technology, but is using the creaky old device that has been left on the shelf and devising a new aesthetic practice for it. He is not creating a nostalgic picture of an outdated and obsolete technology, but rather is offering a reinterpretation of the technology.Through his various techniques Marclay is using the technology in a generative way, forcing it to do something new.

Marclay is critical of digital technology as it is less “hands on”; he argues that CD technology has limitations and cannot be used as a live performance tool. These are valid observations, but the conflict here is that Marclay, as we have seen, is interested in exposing the limitations of phonographic technology, and he exploits them aesthetically. Like CD technology, the phonograph has many limitations, but these are not an issue for Marclay; instead he utilizes and exploits these limitations in performance. As we will see in later sections, the CD and digital technologies in general have been used both in terms of catastrophic failure and breakdown, as well as for the deliberate and forced error.

While Marclay is a dominant and key presence in his practice with the record object, a number of other practitioners were experimenting in similar ways, and numerous subsequent artists and musicians have been influenced by his work. Two such examples are Gum and Vicky Browne. Both show similarities with elements of Marclay’s practice and also demonstrate the multiplicity of techniques in record experimentation.

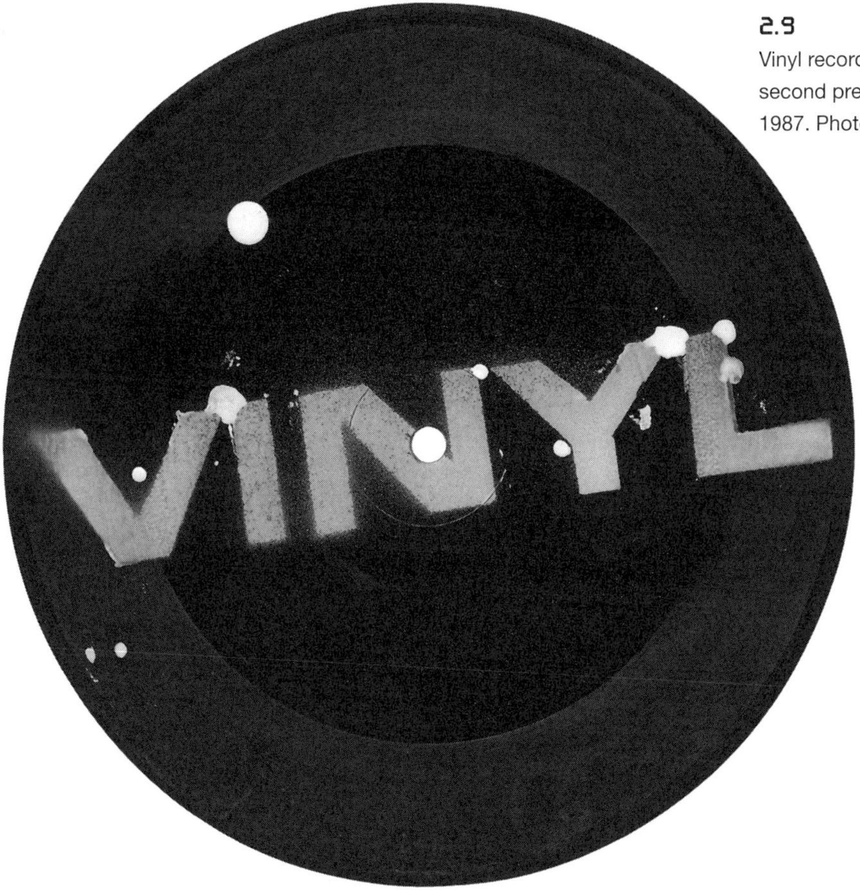

2.9
Vinyl record that was affixed to the second pressing of Gum's LP *Vinyl*, 1987. Photograph: Andrew Curtis.

In a later but similar fashion to The Bachelors, Even, the Melbourne, Australia–based duo Gum (Andrew Curtis and Philip Samartzis) began in 1986 to perform using turntables, multitrack tape, and effects boxes.[132] The band experimented with an array of destructive actions on vinyl records that were sourced cheaply from secondhand shops for a few cents each. These records did not get off lightly in the hands of the industrial music–inspired band. Like Thomas Brinkmann much later, they originally cut lines into their records, creating a rhythmic ticking texture when they where played. This led Gum to sand their records to varying degrees, to the completely destructive act of melting the discs in the oven. Once the discs were melted they would embed objects, such as pebbles and metal filings, into them. The discs where played on old secondhand record players, many of which had variable speeds. At the time Gum had not heard of the vinyl experimentations of Marclay, nor the work of Knížák; instead like Marclay they were influenced by the punk ethos and industrial music, as well as the rhythmic and looping compositions of Steve Reich. Sydney bands Severed Heads and the Loop Orchestra, with their use of repetition, were also factors in Gum's development. But for Samartzis it was the "degradation and the residue of process" that interested him in the use of vinyl records.[133]

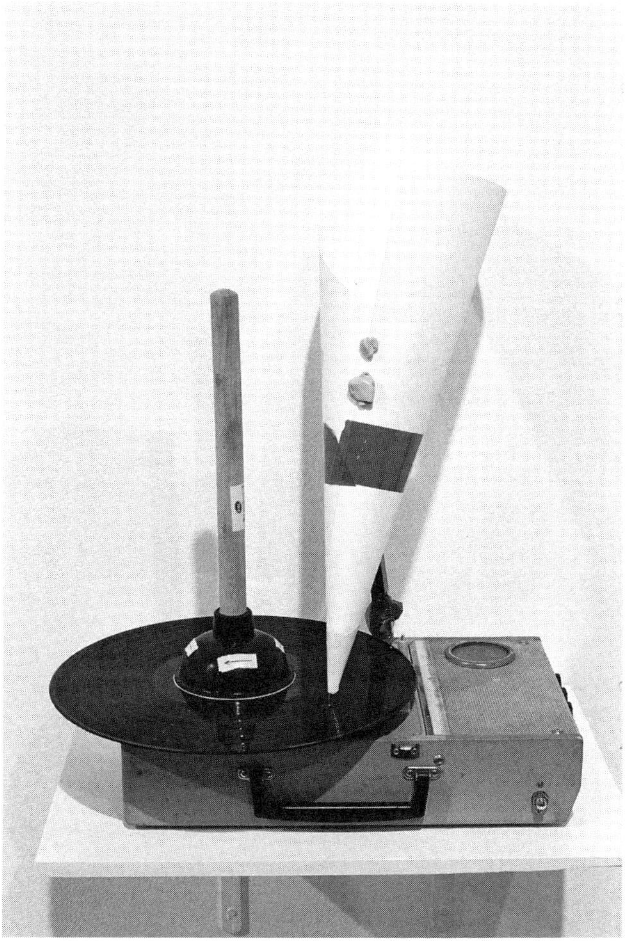

2.10

Vicky Browne, *Work Hard/ Play Hard.* From the exhibition *Mistral*, Artspace, Sydney, 2006. Photograph: Mr. Snow.

Vicky Browne, a sound artist from New Zealand, produces exhibitions that use the turntable as the most dominant element. Her exhibitions focus on the trickery of sound. At Sydney's Performance Space, as part of the 2004 "i.audio" event, Browne's installation *I Want to Be at Home and Feeling Right* (2004) included a copy of Led Zeppelin's *IV* that had been produced by applying glue to the original surface of the vinyl LP. This created a copy of the album that played in reverse, a reference to the alleged practice of hiding messages on a record called "backmasking" that can only be heard by backspinning the album. In the show "Does the Sound of Your Own Voice Soothe You" (2005) at MOP gallery in Sydney, Browne cut a copy of Metallica's *Black Album* into many slices, reforming the disc and allowing the audience to play it on a cheap turntable. In addition she had also produced a cover for the album made up of multiple cuts collaged together. In another work in the same exhibition Browne produced a turntable that had its cartridge head replaced by a tape head pickup. This modified cartridge was placed on a "record" produced from multiple strands of magnetic tape. Browne's work directly references the work of Knížák, Marclay, Nam June Paik, and Laurie Anderson, but from a 1990s popular cultural position.

These two examples are significant, as they reflect two distinct practices that make use of the record itself. The destructive vinyl performances of Gum demonstrates that the use of broken vinyl occurred independently from the lineage of Fluxus, Knížák, and Marclay. Instead, Gum is more closely related to underground popular musical approaches to experimentation, and they have their historical lineage located in the work of industrial rock's experimentations in the 1980s. Browne's practice with records shows how the techniques used since Knížák have been maintained within contemporary art practices and demonstrates how they have developed within a contemporary context. These directions are distinct from the practices of the musicians discussed in the next section, who turn away from the record and instead crack and break the turntable itself.

Christian Marclay, in his destruction of the vinyl object, heralded the next step in the use of turntable as an instrument, which was the disappearance of the vinyl record altogether. In more recent practice decks are used without records at all; the disc is replaced by the rough audio teased or ripped out of modified and manipulated turntables. In the work of Otomo Yoshihide, Akiyama Tetsuji, and Michael Graeve, attention is turned toward the hardware of the phonograph itself. The next section charts the latest stage in the expansion and destruction of the turntable as an instrument.

6 Destroyed Turntables at the End of Gramophonic Culture

Introduction

At the What Is Music? festival in Melbourne in 2002, Otomo Yoshihide played two highly amplified turntables in front of a sold-out audience.[134] The performance was initiated by Otomo slamming down a small cymbal on the tone arm of one of the turntables, and then, while holding the cymbal over the cartridge, seemingly grinding it into the deck. The action caused a visceral assault of feedback to scream through the PA system. Throughout the performance he attacked the two turntables, bashing them with drumsticks and his hands, throwing cymbals down onto the platter and cartridge, as well as vinyl records—though these were never played, as there was no stylus in the turntable's cartridge head. The tone arm was thrown about, the turntable deck left to rotate, shifting and banging the tone arm against the cymbals. At times Otomo allowed periods of simple and constant feedback to play out before spewing forth an array of screeches and high-pitched audio. A complex feedback loop was created by the cymbal being placed over the cartridge head, which had itself been prepared with contact microphones replacing the stylus. The original sound of the cymbal crashed onto the prepared cartridge head was picked up and played back through a guitar amplifier placed near the turntables and turned to extremely high volume. The sound produced by the guitar amplifier resonated the cymbal itself, which in turn was picked up by the contact microphone, causing a feedback loop. The sounds were simple in their use of extreme volume and noise, yet also full of layers of texture, some delicate and fragile. The audio can be likened to guitar feedback commonly used as an effect in rock music, except Otomo's set was solely made up of these sounds, at once familiar and ear shattering. The sound was extremely loud and painful; paint crumbled off the ceiling of the Melbourne rock venue Revolver. The audience was surprised by the perceived violence of this performance, in terms of both volume and Otomo's actions. In fact the seeming violence of his actions caused no actual damage, as he had earlier replaced the styluses of the turntables with contact microphones inserted into the cartridge heads.

Whereas Christian Marclay focused predominantly on the record disc, in the second half of the 1990s the use of the turntable as an instrument moved away from playing records, prepared or otherwise.

The vinyl album, with its grooves of recorded information or its artistically inscribed marks had, after all, already been destroyed. The next step was to destroy the turntable itself. What occurred was a type of phonographic cleansing, as all external input was removed, leaving the technology to play its own death rattle. Vinyl seemed to have been forgotten by the next wave of turntable experimentalists. Here the musicians were exploring, not the possibilities of vinyl and its inscribed data, but rather the playback technology itself. In the process they discovered the boundaries and edges of the phonograph and worked with its flaws and faults. In turn the sound and music they produced was completely unlike either the dexterous vinyl manipulation of the hip-hop turntablist or the radical juxtaposition of sound bites in Marclay's production.

The devaluation of vinyl had led to the near worthlessness of the turntable itself.[135] The phonograph had been roughly stacked in the back of secondhand dealer shops, discarded and in disrepair with old blunt needles and worthless components. These turntables could not easily be sold, and more importantly they could not be fixed, as parts for replacement were no longer available. If turntables were cheap and easily accessible, then they could be easily ruined without serious financial loss. These decrepit record players were laid open for destructive experimentation and the devices were modified, cracked, and broken in the process.

The discussion here shifts from preparing and extending the information stored on the disc to preparing the turntable itself for performance. Here the needle is forced to read its own technological support structure rather than the grooved surface of the disc for which was designed. The needle no longer translates vibrations into recognizable music but rather outputs audio filled with scrapes, rasps, earth hum, and feedback. The sound is quite simply noise. There is no reference, no playback of a recorded moment from the past. An internal feedback loop is created as the needle reads its own body, playing itself back onto itself. Many of the sounds in this practice are produced through the use of feedback, as minute sounds are heavily amplified and in turn vibrated back through their own amplification.

Starting with Otomo Yoshihide, this section will follow the extremes of his practice and a number of musicians near him using turntables in the late 1990s and early 2000s. This will include the turntable's amplified motor hum in the work of Martin Tétreault, the drone-based practices of Michael Graeve, and the extreme modification of the phonograph by Lucas Abela. This section is by no means a complete survey of the area but rather a focused study of a small number of similar practices.

Otomo Yoshihide and the Turntable without Vinyl

Otomo Yoshihide is an influential and significant part of the Japanese experimental music scene.[136] He emerged in the early 1980s as a guitarist from the free jazz and free noise scene, influenced by Abe Kaoru and Takayanagi Masayuki, before becoming known as an experimental turntable player. Originally he utilized the turntable in his performances as a sampler, pulling references explicitly from various sources in a manner heavily influenced by John Zorn. The extreme use of samples, especially the approach taken in his band Ground Zero in albums such as *Consume Red* (Sank-ohso/Creativeman, 1997) and *Last Concert* (Valve/Amoebic, 1999), has been replaced by an instrument without external memory. Since the late 1990 Otomo has turned away from playing records on his turntables.

With the recordless turntable Otomo coaxes out abstract sounds. Very small sounds, clicks, scrapes and rattles, that are initially very quiet, are heavily amplified. He explains:

> I am interested in the fact that people have started using the turntable as a musical instrument—recycling it, in a way—

> just when its role as a practical piece of audio equipment was ending and it was about to retire from the scene. In this sense I'm interested in the work of musicians like Christian Marclay, Martin Tétreault, Philip Jeck, and Tetuzi Akiyama. Yasunao Tone's way of using CDs is extremely interesting, too.[137]

Clearly these thoughts are directly linked to the belief that the destructive use of the phonograph has followed the slide of technology's commodity value. Otomo talks of the device being recycled. The turntable is given a second chance, pulled out of the rubbish and reconsidered. Its very obsolescence as the playback device of choice allows for the technology to be modified and broken, rethought of beyond simply being a playback device.

Otomo's work as a noise musician came to the fore in his band Ground Zero, which was created to perform the John Zorn game piece *Cobra*. The band first performed in August 1990 and had its final performance in March 1998.[138] Ground Zero was a large group and many different musicians played in it over the seven years it existed, with the final performance involving thirteen musicians. Its sound was loud and extreme, a virtuosic mix of free jazz, improvisation, rock, pop, and experimental noise. In line

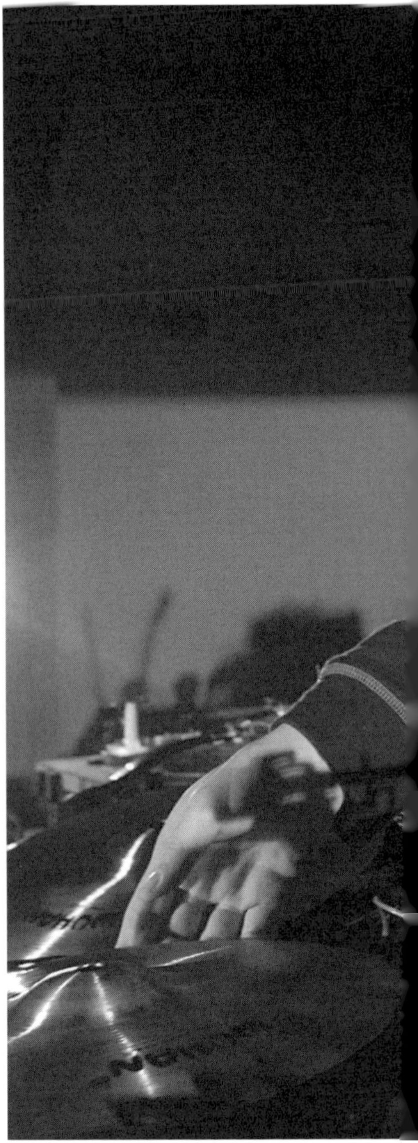

2.11
Otomo Yoshihide in performance, 2007.

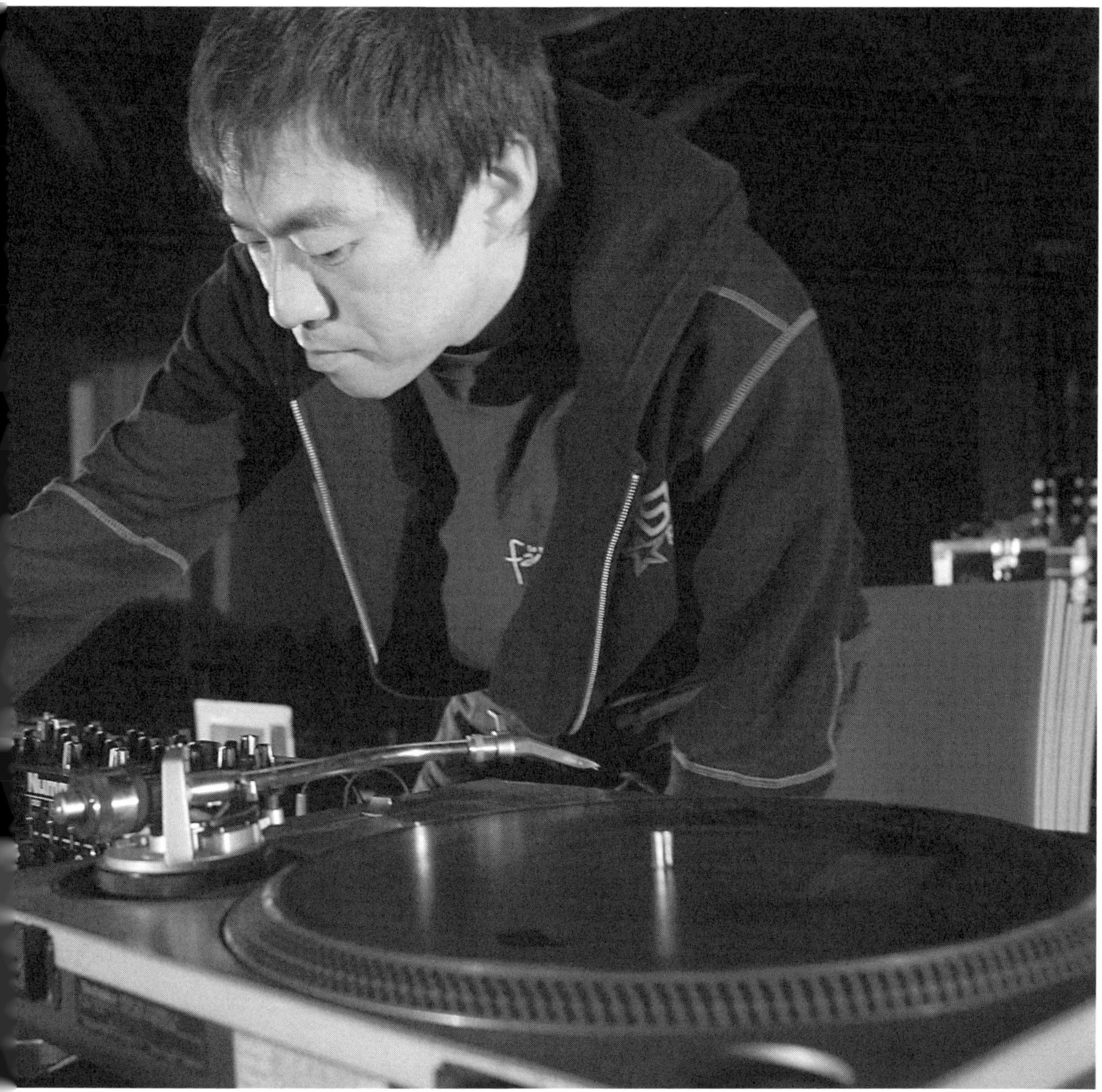

with John Zorn's approach, the band sought technical mastery of their instruments, including Otomo's use of a turntable as a radical sampling machine.

Otomo's later break with the style of performance of Ground Zero was equally extreme, as he shifted away from using his turntable as a sampling device toward using it as a vinyl-less, abstract noise generator. At times he played the turntables exceedingly quietly and with restraint, and at others with a shocking and violent performativity. Otomo has said that he was bored with producing music that made its effect through sampling and the memories associated with it.[139] He explains:

> Before I was mixing a lot of information together, but now the real world is just like that, especially in Japan. People are barely surviving in a sea of too much information. . . . Anyone can sample cheaply and easily now. So now I want to focus, to find a whole world in tiny things.[140]

Part of this reasoning was the feeling of absolute control he had over the sampling technology and his turntablism. An extremely virtuosic musician, Otomo explains:

> I quickly get tired of things that I can control myself. I started to think that using the turntable without records and making feedback, gathering noises . . . that this kind of process, which is much less controllable than sampling, was extremely interesting. I discovered specific techniques by actually working with I.S.O. and Filament.[141]

As described at the start of the section, one of Otomo's performance styles is extremely physical and active and can be likened to a rock guitar solo. At the other end of the spectrum is the duo Filament (Sachiko M and Otomo) who perform in completely opposite manner. At a Filament performance in Yokohama, September 29, 2000, Otomo used a prepared cartridge made from a piece of metal soldered onto the tone arm. The metal had a spring on one end and was bent at a right angle. It sat on the turntable's platter (without a slip mat), the modified turntable amplifying the sound of metal being drawn across metal as the platter spins. This prepared turntable created a high-pitched tone that was similar in effect to the sine waves being generated by Sachiko M's "emptied" sampler, but with a more metallic resonance.[142] The sound space of the theater in Yokohama was filled with gentle, quiet, sine-tone-like sounds.

Otomo's turntable has been turned into a noisemaker, a physical platform from which sounds are generated. This approach can be clearly heard in a duo with Akiyama Tetsuji (known in the West as Tetuzi Akiyama) recorded at the gallery and venue Off-Site, in Tokyo, in August 2002. The piece is an improvisation for two turntable performers, and the recording is rich with both amplified and acoustic sounds (it was recorded with a live microphone in a venue without the use of a PA, allowing for the acoustic thumps and bumps to be clearly heard). The turntable is fully utilized as a physical noisemaker in this work. Extremely quiet sounds are amplified: quiet bangs, ticks, squeaks, creaks, and groans. There is nothing about these sounds that has the signature of a record player—usually a combination of the grain of the record and the music stored on the disc. Also, the turntable is not usually played as an acoustic instrument. After Otomo, it is possible to imagine a completely acoustic performance with a turntable in which no amplification occurs. In this situation it is the housing, the deck, and the rotating platter that are played to produce sound.

Akiyama and Otomo were part of the "Onkyo" sound scene in Tokyo. For a time many of these performances occurred at the gallery Off-Site, situated in Yoyogi under one of the largest skyscrapers near Shinjuku in Tokyo.[143] Off-Site's location is an important factor in

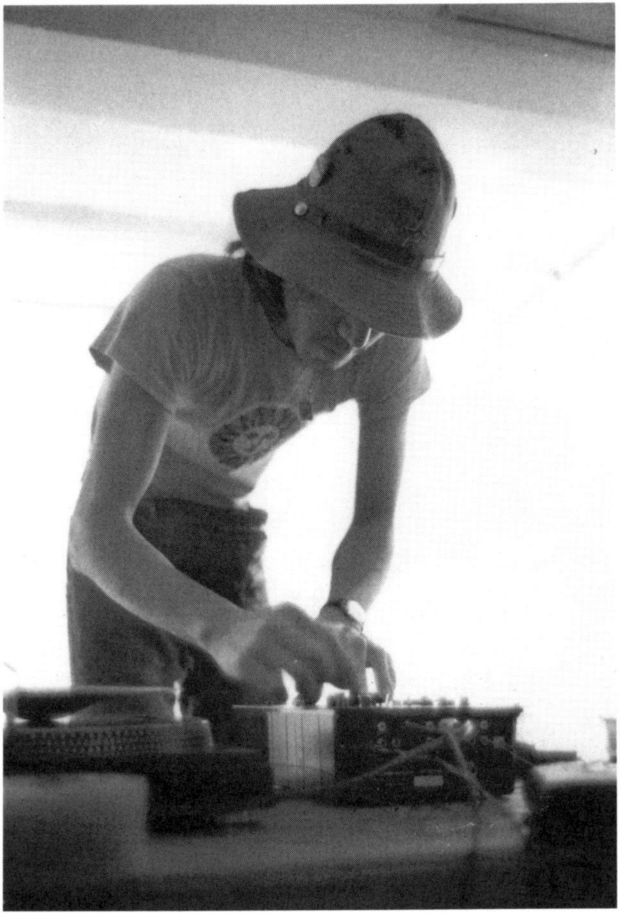

2.12
Akiyama Tetsuji, performance at Gendai Heights, Tokyo, 2000. Photograph: tanabemse.

the development of the music that was performed there: any music played needed to be quiet so as not to disturb the neighbors.

Akiyama performs mostly with guitar but also utilizes turntables, electronics, viola, and self-made instruments. He developed his improvisational style using a guitar prepared and extended with various tools, including notoriously a family samurai sword, to coax an array of unique sounds out of his instrument.

In May 2001 at Off-Site, Akiyama, performing with a turntable, joined a group for two improvisations with Nakamura Toshimaru, Yasunaga Tetsuro, and Tim Olive. At the center of an array of electronic equipment was a Technics turntable. To its direct left was a DJ-mixer, though this set would bear little resemblance to the Technics turntable and mixer performances of the dance turntablist. Attached to Akiyama's turntable were various contact microphones, placed on and around the deck, one attached to a plastic clip. In addition to the contact microphones Akiyama had an old CB-radio hand piece, which was placed near the turntable. During the two sets of the night Akiyama, like Otomo, used the turntable more as a "noise generator" than as a playback device.

The techniques that were brought into play included placing the contact microphone attached to the clip onto a cymbal, which was itself on the deck of the turntable. The cymbal was then activated with various objects. A wire spring was strung across it, and this was rubbed with various objects; sugar was poured over it, compressed air sprayed at it. The cymbal was removed and the turntable's platter was set spinning, the dimples on the side of the spinning deck bumping the CB-radio handset. The slip mat from the platter was removed and a chain of metal balls was left to spin off it. The needle of the turntable was also used. It was placed on the cymbal and allowed to thump down, and was also placed on the slip mat before the platter was spun, causing a rasping rumble. Here the needle simply became a contact microphone. Akiyama has gone well beyond the idea that the turntable can be an instrument, which he takes for granted; he has shifted to using it as an abstract sound-creation device.

As with Otomo's technique, Akiyama's approach does not actually damage the turntable. There is damage to the usual workings of the instrument in terms of its technical features and intended use, but this leaves no lasting marks. Instead Akiyama's use of the device can be read as a manipulation of the mechanics of the tool. Having removed the software, he has extended its mechanisms. The slip mat has been taken off the platter, external objects have been placed on it and nearby, elements such as the dimples on the side of the plate are utilized, and the needle has become a contact microphone. The turntable is an abstract noise generator from which Akiyama pulls thumps, whirs, ticks, pops, hisses, clicks, and rasps

Akiyama's approach to his instruments can be likened to British improviser Keith Rowe's use of the guitar as noisemaker in performance. Gary Butler explains that:

> The notion that Rowe has constructed a new instrument in which the guitar is merely one of the components is a useful idea. However, as Rowe's use of preparations constantly changes during an improvisation, it is more useful to think of this as a series of constructions, with the electric guitar featuring as a common element included in each new version.[144]

Similarly, here the turntable starts to disappear, hidden among a collection of electronic equipment. It is not the focus of the performance in the eyes of these musicians but rather another element in the construction of an improvised performance piece. The link to Keith Rowe's performance is useful here, and we might compare his guitar, which has a very small body and as such can fit and blend in with an assortment of equipment more easily than a conventional electric guitar, to Akiyama's turntable. In Akiyama's performance the turntable also begins to blend in, especially as we become familiar with its extended use. The turntable in the hands of Akiyama and Otomo has also been played purely acoustically, with no amplification whatsoever. In this case we have a technology that resembles the prepared and extended use of the acoustic guitar.

The Rumble and Rasp: Michael Graeve and Lucas Abela

Each individual turntable can be heard to have different characteristics and sound qualities. This perception was voiced by Philip Samartzis, who described all-in-one phonographs (the type that have their own amplifier and speakers built in) as having their own unique sound.[145] Given this individuality, specific instruments can be used for specific effect. This is the case for Melbourne-based artist Michael Graeve, who uses multiple phonographs in performance, each with its own amplification. Graeve does not use a PA and as such requires a large number of phonographs to produce enough sound to fill a space. In performance he uses domestic all-in-one stereo equipment with up to twenty-five players and fifty speakers being utilized in a single performance.

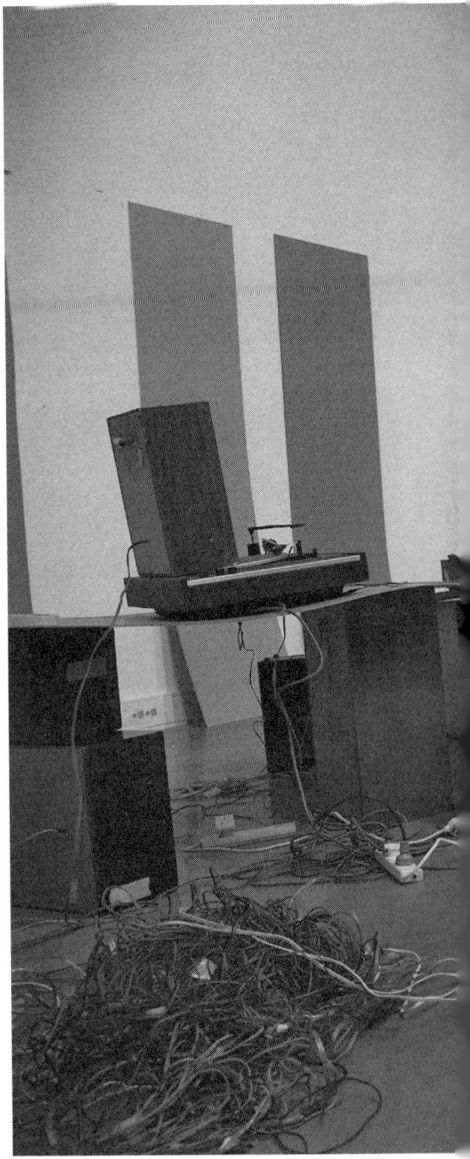

2.13

Michael Graeve, untitled performance and installation at Monash University, Melbourne, August 2007. Photograph: Christian Capurro.

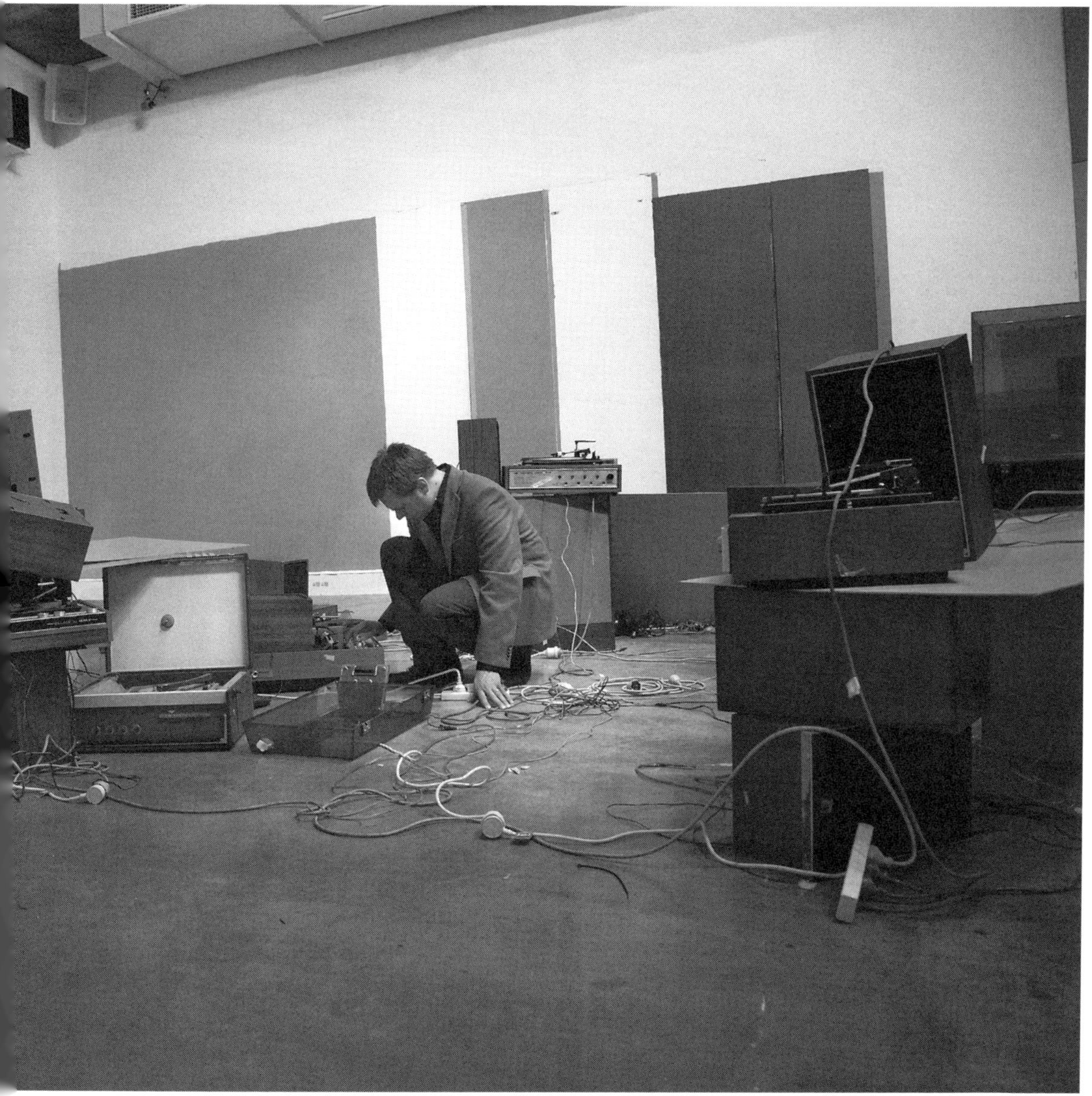

Graeve writes that he "assist[s] these systems in picking themselves up: Rich textures and rhythms fall together and fall apart, evidencing simple interactions between machine process and human gesture."[146] In performance he employs a number of techniques including; turning the player on and off, with accompanying machine sound or lack thereof; changing the volume levels; increasing or decreasing the bass and treble; shifting the balance; speaker placement in the room; quality of speakers (or lack thereof); turntable engine sound; player speeds and pauses (nearly all the players have 16, 33, 45, 78); a variety of platter/mat surfaces, or taking the mat off completely; quality of needle, or lack of needle; and placement of needle (on metal the stylus often transmits engine hum or high-pitched feedback, on a rubber surface it gives a dull feedback, and on wooden surfaces it gives a low feedback).[147] These techniques are contrived to create different sections of sound in performance. For example, the most obvious method simply involves placing the needle down on the rotating platter and adjusting the speed and volume levels to produce an array of hisses, rasps, and rumbles. In addition, Graeve deliberately amplifies the volume level of the turntable's motor by placing the needle close to it, treating the needle once again as a contact microphone, creating a hum rather than the grainy rasp of the needle on the platter. He also uses feedback as a device in performance, placing the speakers in the vicinity of the needle. The speaker sound is picked up by the needle itself, creating a feedback loop. In performance he might move the speakers, or lean them against the record players, to manipulate this feedback. These sounds can also be altered by changing the volume, allowing a little through at a time, or letting loose into a loud section of feedback. Graeve also prepares his turntables by placing masking tape on the platter. Over the course of numerous performances the record players themselves have worn down, deadening the edge they originally had. Early on, the tape created a rhythmic tick, but now it often simply sticks to the needle. He also places pegs on the platter so that the needle is kept in place or regularly hits the peg itself.[148]

There is a striking physicality in Graeve's performances as he uses the performance space in the manner of a visual installation rather than a traditional concert or performance setting. The players and speakers are often scattered around the room and he must move about his equipment to make changes to individual players. In most performances he does not use any PA, so it is not simply a matter of tweaking the mixing-desk to effect changes in the audio. When watching Graeve in performance, it seems that he is perhaps closer to the DJ or mixer than a musician, in the sense that often he places the needle onto the deck and lets it run its course with little or no intervention. There is plenty of time for him to contemplate the proceedings before he scuttles around his equipment, readying for the next change, whether subtle or dramatic.

For Graeve, however, the weight of the history of the technology itself plays an active role in the way in which the audience comprehends the performance. He explains that the record player comes with its own cultural baggage and that there is an expectation from the audience of what this technology does. Yet at a performance by Graeve the tool is forced into playing audio that we do not expect; there is a "disjuncture between knowledge and experience."[149]

Lucas Abela's Dangerous Phonograph Devices

Sydney-based Lucas Abela has been modifying turntables in dramatic fashion since the early 1990s, and in the process he has become one of Australia's most respected noise musicians. His practices are based around building instruments, many of which utilize the turntable at the core of their mechanics. He has become best known for his project entitled *Justice Yeldman and His Amazing*

Ribbon Device, for which he blows onto a pane of glass with a contact microphone attached to it. The amplified sounds created by Abela are in turn processed through guitar effects pedals and emitted through a PA at high volume. Abela risks injury during the performance as there are no safety precautions taken with the sheet of glass. He often nicks, cuts, slices, and gashes his face during performances of the work. Although this work is directly related to his earlier extended turntable works, it no longer uses turntable parts.

Abela initially began experimenting with records after hearing about industrial music maker Boyd Rice, who has released vinyl records with a second off-center hole in the middle that allows for a skewed playback of the record.[150] The majority of Abela's work is based on the extension of the turntable's cartridge head and the reinvention of the turntable platter. An instrument called "The Bridge," for example, was created out of a range of records, or record substitutes in the form of found disc-shaped objects, attached to a series of engines that spin the discs at high speed. Instead of a regular needle Abela used foreign objects such as springs, swords and skewers as makeshift needles. As might be imagined this instrument was highly risky to use; after one public performance with it he was hospitalized with a severed radial nerve.

Less extreme instruments he has built make use of different types of needles, objects that are engaged on records, various types of discs, and sometimes body parts. One instrument known as "the Humming Bar" is simply a skewer attached to a turntable cartridge that is played through an array of effects boxes strapped to Abela's waist. The bar is placed in his mouth and he produces sound by running the bar across his teeth and grinding it against his tongue. This instrument is similar to the one that utilized an industrial sewing machine (described at the beginning of the book), in that the skewer forms the stylus of the extended turntable. A less extreme instrument was made out of a squash ball cut in half with many pins pushed through the rubber of the ball. On the back was a contact microphone attached to a phonograph. When placed on the surface of a vinyl disc it had the ability to pick up multiple grooves from a record at the same time. This multiple-needle device was often used in conjunction with house music records with the aim of creating out-of-sync sound from the simultaneous playback of multiple grooves of beat-driven dance music. The cartridge is hand-held; this is certainly not a precision device, and it tends to work by pushing and tearing the pins across the vinyl surface.

Abela's work raises the question as to whether these instruments should still be considered turntables. They retain elements of the turntable, and his practice continually returns to the phonograph and vinyl. The almost unrecognizable phonograph has been disassembled, some of its pieces kept while others are discarded; yet there remains a phonographic approach to sound making through these last remnants of the machine.

In many of the practices described above the turntable has been used as a generative sound device and the sounds it creates are well beyond those originally intended for the technology. Perhaps the most obviously generative practices come from these turntable musicians who pull noises from their turntables without the use of vinyl. Here the noise of the technology itself generates sound and a feedback loop is produced as the device plays itself. In the work of Michael Graeve this occurs through an orchestra of turntables set loose on themselves, whereas musicians such as Akiyama use the devices as noisemakers in conjunction with an array of other elements and devices. These turntables do not resemble the original technology at all; the needle is almost never used and PA amplification is drastically reduced or abandoned altogether.

A key element of almost all of these turntable practices is the allowance for noncontrol over the sound outcomes of the work. An indeterminacy enters the work, as the exact nature of any sound produced is not known in advance by the artist. Artists seek out uncontrolled sound and compositional structures by utilizing the generative possibilities of the noisy, cracked, and broken turntable. This approach was discussed in relation to Marclay, who saw exciting creative potential in the unpredictability of the unintended needle skip.[151] In the hands of musicians such as Otomo, this lack of control creates a chaotic performance in which errors or accidents are no longer perceivable among the array of noise. The musicians discussed in this chapter are largely involved in improvised music, a form that often allows for and desires the unknown, the uncontrolled, and the novel. Within this framework, one of a desired aesthetic of error, mistakes and noise become positive elements of the work.

3 damaged sound

glitching and skipping compact discs in the audio of yasunao tone, nicolas collins, oval, and disc

1 The Sound of Glitching CDs

A common experience for any regular café goer is hearing the sound of a skipping CD.[1] Café music is most often intended for the background; the music is not there to be listened to in any attentive way and is most often ignored until the CD, perhaps covered in scratches and dust and being played through a second-rate stereo, inevitably starts skipping. The stuttering CD in the café environment causes all to stop and take notice. The high-pitched glitch of a CD "hanging" on a small section of audio causes an instant reaction. The customers look around for someone to fix the problem, by forwarding to the next track or changing the offending CD altogether, so that they can go back to their conversations. What could be worse than the sound of a skipping, glitching CD?

This section will examine in detail two individual producers and two groups who create works based on the damaged sound of the distressed CD player as it tries to cope with the loss of binary information. Yasunao Tone and Nicolas Collins and the groups Oval and Disc heard these digital sounds as ripe for exploitation toward sound expansion, indeterminate composition, and even pop music. The different approaches of the four are discussed in detail: Tone's interest in chance and the singularity of the performative situation; Collins's modification of the CD player itself; Oval's sampling and sequencing of the glitch to create abstract pop music; and Disc's extreme abuse of the CD.

The use of CD skipping in music has been an extremely productive approach and has itself, especially through the popularity of the music of Oval, spawned a subgenre involving what Collins calls the "benevolent catastrophe."[2] Tone, Collins, Oval, and Disc, however, have divergent approaches and interests. These interests stem from the musicians' own very different histories. For example, Tone was involved in the Tokyo Fluxus movement and has a long-standing interest in indeterminate composition. Whereas Collins, like Tone, is involved in avant-garde work, his history is closely tied to the experimental electronic practices of the 1960s and 1970s in the United States. Both Collins and Tone have been involved in the contemporary digital audio scene. Collins remixed a track for Microstoria's *Reprovisors* (Thrill Jockey, 1997), a side project of Oval's Markus Popp; in 2002 Tone won the Golden Nica for Digital Musics at the Prix Ars Electronica. Oval, on the other hand, developed out of an indie pop background and subsequently crossed the boundaries between art and pop. Similarly, Disc was spawned by the "DIY" digital production culture that later took up the technique of the crack, and combined a punk aesthetic with electronica. Despite this diversity, all of these producers have in common the practice of cracking, breaking, and transforming the mediated sound of the CD system and, through various manipulations, engage in a practice that crosses between music, sound, and noise.

The CD system is a relatively new technology. It was initially released on to the market in late 1982 and early 1983, and as such its release and subsequent development has occurred within living memory, unlike phonographic technology. Whereas the phonograph and its experimental practices have developed for well over a century, the CD has a short history of some twenty-five years. This short period is indicative of the increasing rate of technological change; new technology quickly takes hold of the market and older technologies just as quickly become obsolete. The CD by the 1990s was a part of our daily life. CD hardware is now ubiquitous, appearing in lounge rooms, portable devices, and computers; it is used for its original purpose, playing music, as well as for data storage and video playback, given the subsequent development of the digital video disc system (DVD).

The possibilities of the CD system for experimentation in music were taken up extremely quickly. Within a year of the technology's release, Tone was developing a method to make use of the flaws in the system for the purpose of performance, whereas phonographic experimentation was relatively slow to develop, with some eighty years between its initial invention and John Cage's *Cartridge Music* in 1960.

Artists using CD technology have further developed the practice of cracked media in a number of ways, taking off from the turntable experimentation of Christian Marclay and redirecting the crack and break down a new digital pathway. Marclay is closest in intention and output to those who took up the system, and the links between Marclay and Tone especially are clear. Marclay prepares his vinyl records for performance by sticking foreign objects on the surface of vinyl discs, causing the needle to skip and jump across the surface. Tone did something very similar for his "Wounded CDs," placing Scotch tape on the substrate of CDs to force them to skip, stutter, and glitch. These practices also have a visual similarity in that the surface of an object, which is typically meant to be kept clean and protected from damage, has been marked. Both musicians work with the inherent noise in their system, as the broken flow of data, caused by the manipulated surface of the disc, is converted through the mediation process to noisy audio.

There are also broader parallels between the practices of cracked turntables and CDs. The most obvious is an interest in the use of a technology beyond its design and intention. The technology, designed for the mediation of recorded information and its playback, is pushed and pulled, tampered with, bent, cracked, and finally broken. In both practices, the playback technology is redirected to play the “noise” caused by, for example, surface disturbances, as content. That is, the device is no longer playing recorded material but instead is focused on physical cracks and breaks in the surface caused by the attachment of foreign objects or marked into the surface in the form of scratches and cuts. Whereas the specifics of the technologies, as discussed below, and audio outcomes are very different, both practices use everyday technologies to produce unique audio outcomes, rerouting the system and exploiting a stressed medium for the generation of new sounds and compositional practices.

This redirection from signal to noise can also be understood as an amplification of already existing flaws in the media. It is the defects or errors in both systems that are employed, those sounds that are not meant to be part of the listening experience. For the phonograph, it is the pops and ticks of small scratches and dust; for the CD it is the skip and stutter, initially caused in the same way. The seamless, high-fidelity listening experience (the very reason the CD player and its digital audio mediation were developed) is broken by these composers who seek to find new sounds and expanded practices in the systems' cracks and breaks.

The glitching CD of Tone and subsequent experimentations with the technology do, however, diverge from the use of the turntable and add to the practice of cracked media in a number of important ways. First, the turntable is an analog reproduction device, whereas CD technology is digital. The very physical nature of the cracked and broken phonographic mediating device is inherent in its exploitation for experimental sound practices—recall Otomo Yoshihide's aggressive performances with the Technics turntable. The CD is not open to such a physical approach; a laser cannot be roughly handled, unlike a phonograph's tone arm, and unlike the turntable, the CD player cannot be forced to continue to output audio once the CD has been removed. The direction taken by many no-input turntablists and their non-use of vinyl, is simply not an option for those using CD technology.

Furthermore, it is the very *clean* nature of digital audio that is highlighted in the use of CD technology. Vinyl is a very tactile medium; its users can literally get themselves dirty, handling the vinyl record and turning it into a sculptural form in the cases of Milan Knížák and Marclay. This connects the practice of cracked phonograph media to a visual arts tradition, quite unlike the extremely clean and fragile environment of the digital studio. Instead, the practitioner of the glitching CD must handle the technology with care. It is seemingly far too easy to cause a CD to skip and stutter; it appears at times that the slightest scratch causes the CD to skip just at the best part of one's favorite track. The irony is that the deliberate forcing of a CD to skip is a very delicate operation. Those who have tried find that the mark must be placed just so on the substrate of the disc, or the CD error correction will either completely conceal the damaged data or simply not play the disc at all. One must approach the destruction of data on the disc with great care.

Another key difference is that CD technology has yet to be truly superseded as a playback format for music (although MP3 technology is perhaps now becoming the playback format of choice); when Tone took to the medium he was using a state-of-the-art technology. This places the CD in a different position historically from that of the later use of cracked vinyl and turntables. Turntable experimentalists such as Marclay used discs that had become a surplus commodity and subsequently a superseded technology; Otomo used the turntable after the death of vinyl and the devaluing of the turntable itself. For the original practitioners of the CD glitch, the format was relatively new and was certainly not an outdated technology. It was not worthless and was certainly not inexpensive, especially when Tone and Collins began working with the device. This means that less extreme and less permanent manipulation took place, and in case of Tone, Collins, and Oval, the objects could be fully restored to their original clean state: the CD could be returned undamaged to the library whence it came.

The digital format of the CD also opened up a new possibility for cracked media, that of the planned use of an accident's trace. Whereas Tone and Collins are interested in indeterminate compositional practices and Collins himself expressly looks for the "benevolent catastrophe," Oval, once they have found the accidental sounds of the CD skip and stutter, record them to use again and again in a highly controlled digital sequencing environment. This is very different from earlier practices with vinyl and turntables: although the practitioners knew that something would happen when the needle hit the scratch on the vinyl disc, they did not know exactly what would happen. They could not know if it would simply "pop" or if it would skip a groove or even skid across the surface. Oval knew when playing a file during a live performance exactly how it would sound; it was a prerecorded and sequenced event that had been placed just so, to create a track more akin to a pop song than the sounds that have thus far been discussed in relation to experimental music and art practices.

Compact Disc Technology and the Glitch

> Incorrect or missing data from a Compact Disc can result in clicks, pops, dropouts, "skipped grooves," and "stuck records."
>
> —David Ranada[3]

Compact disc technology was developed by the Philips and Sony corporations and introduced in Japan and Europe in October 1982 and in the United States in March 1983.[4] The format quickly emerged as the sound-storage and music-publishing medium of choice.

There is a great difference between the way in which information is read by the analog phonograph and the way it is read by the digital CD player. An LP is made up of grooves through which a stylus moves. The change in horizontal and left and right movement is an analogue of the original audio signal. The depth, width, and its variations produce a vibration that is then transduced into an electrical signal and amplified. As the record is played the vinyl wears out owing to the physical contact of the needle as well as general wear and tear. What is physically happening can easily be seen in the case of the record skip. We can see the scratch, the mark or scuff. When the record is being played we can watch as the needle is pulled over the spinning record. As it hits marks, pops are emitted from the speaker system, and in the case of larger scratches or warps in the vinyl we can see the needle jump, leaving the surface before bouncing back into a groove. This analog system is very tactile. The listener can make a clear connection between an action and a sound.

A CD, on the other hand, has information stored on it in the form of tiny "pits," which are invisible to the naked eye and represent binary ones and zeros. A laser is trained on the spiral of pits and the player reads variations in the reflected light as on and off signals. Nothing except light is touching the disc and as such it will not wear out from actual playing. Hence when we hear a skip or pop from both systems, what actually causes that sound is vastly different. The CD player's laser has no physical contact with the pits, and as such, any scratches or marks are not physically hit by the device. The laser is not drawn across any bumps or marks, it does not hit the dust, it does not jump off a rise in a warped disc. As such the CD player's laser does not actually "stick" in the way a phonograph needle becomes physically locked in a groove. The glitches heard from a CD player—the skips and stutters—are not caused by actual skips; instead what we hear when playing a CD are errors being emitted from the system as audio. Binary values are being read incorrectly; if the level of the error is so great that the error-correction software built in to the digital system is not able to cope, then the system will emit *false* sounds[5]—that is, sounds that are not intended for playback or are not in the recording. For a CD to do this it needs to have been handled very badly. Dust and small scratches, especially across the disc, do not normally affect the audio output; errors are concealed by the system's error-correction software, which corrects misreadings and information losses.

At the time of the release of CD technology numerous accounts of the technology advertised the resilience of the medium; one suggested that instead of being promoted mainly to the audiophile, the new technology could be aimed at the casual, everyday user with slogans such as "The record you can use as a beer mat!" or "At last—the record you *can* play at parties!"[6] It quickly became clear, however, that although the disc is sturdy, it should be handled with care so as to protect it from damage and the eventual glitches that result. The reviewer David Ranada warns:

> You should *not* “play Frisbee” with your Compact Discs. They are not indestructible and, if mishandled, can easily be damaged so as to be unplayable. (Real Frisbees, in any case, fly better, are cheaper, and are less prone to fatal damage when mistreated.)[7]

There are four basic types of errors that occur, representing different levels of damage: fully concealable errors, concealed or interpolated errors, muting errors (all signal is muted), and detracking errors. It is the final type that causes the glitch. The CD carries both audio data and tracking information. The laser-reading component of the CD player is controlled by servo motors, which keep the laser “on track”—aimed at the correct point on the disc. There are usually four individual servo systems. The information driving those servos is embedded in the disc as nonaudio information. Lost tracking data can throw the tracking servos off, or “detrack” them. Extreme damage to the pits can cause this information to be unreadable and produce CD glitches, the result of the servos getting “stuck” and forming a type of locked groove, although the mechanism is not physically stuck in a groove but is instead lost.[8]

Testing the System

Even before the release of CD technology, David Ranada of *Stereo Review* magazine sought ways to test the system for himself. For his first feature on CDs, Ranada worked out a method to make a CD jump and skip. The resultant review was printed in December 1982, before the commercial release of the CD player in the United States. He described the error-correction system and his attempts to cause significant data loss:

> If the maximum burst-error length that can be concealed . . . has been exceeded . . . a CD player stands a good chance of either emitting very strange sounds or muting altogether. And that is just what happened in my listening tests.[9]

Ranada made "a very thin wedge of laser-opaque adhesive tape and affixed it to the surface of the CD, pointing it toward the centre of the disc."[10] As the laser moved across the tape, the CD went through the various types of errors described above until it "stopped and its disc drawer opened, encouraging me to remove the offending record."[11] He compared his most extreme test to sawing out a slice of a vinyl record and trying to play it back on a turntable. In a later

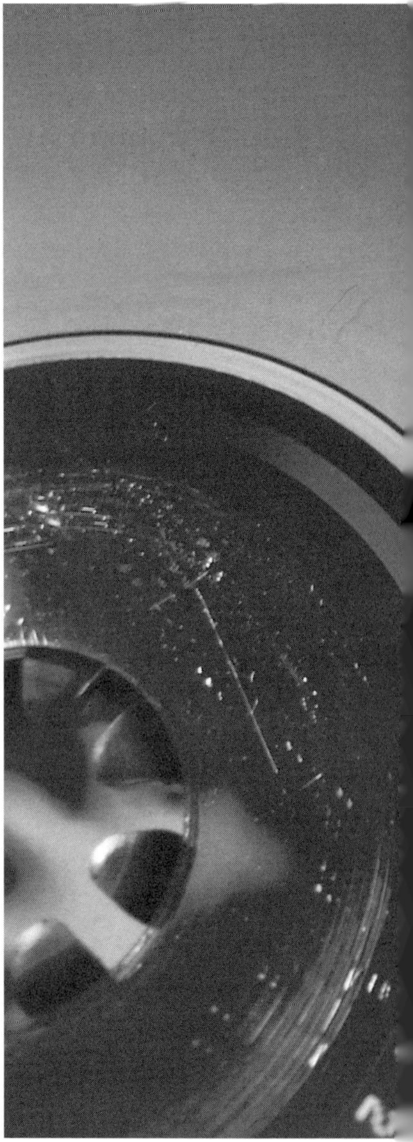

3.1
Philips "Test Sample 4A" CD (410 056-2) with surface and engraved defects. From left to right, interruption in the information layer (made by damaging the stamper), diameter-calibrated black dots on the disc surface, and a simulated fingerprint also on the disc surface. Photograph: David Ranada.

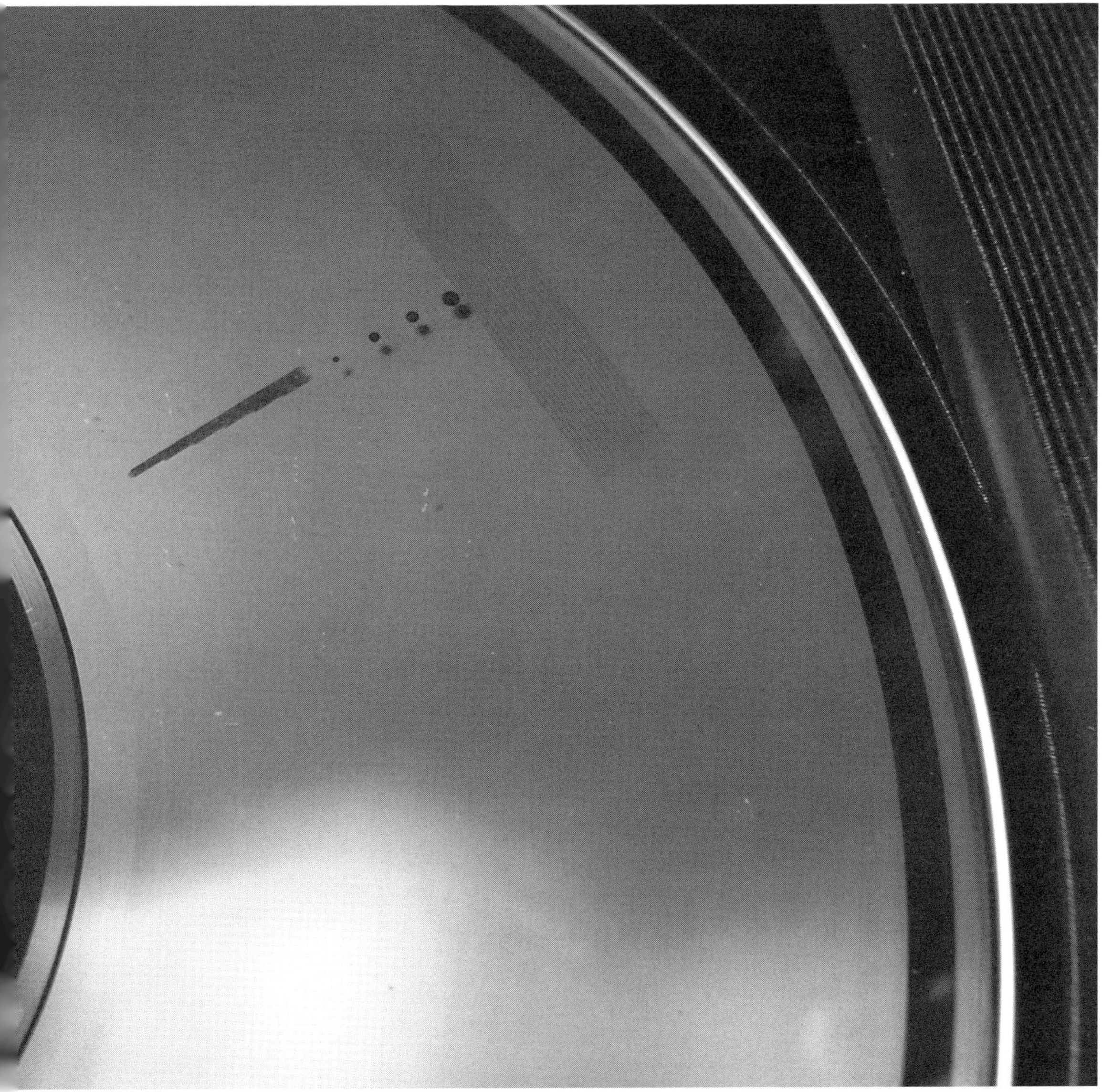

article Ranada notes that the wedge could be made from Scotch Magic Tape and describes in more detail how wide he made it.[12] He found by examining the circuitry that he could read exactly how many errors were being concealed and realized that error correction was itself erratic and would often deal with large errors in different ways from playing to playing. This became important for future uses of the technology, especially for those musicians who push the system in an attempt to draw out the accidental and unforeseen from the device.

What this demonstrates is a common interest between artists and the reviewer who tests a new technology by forcing it to fail. The manufacturers and developers had created a new form of technology that was said to be able to correct both errors and damage done to the discs. In reaction a reviewer sought to find out exactly how good the new system is, how hard will it be to prove it fallible, and what exactly will happen when it does fail. He does not see this process as anything more than a conscientious reviewer doing his job. Nor does he not talk about the actual audio beyond basic accounts of low-level clicks and jumps, which sound to him like a bad slice in a tape recording. The interest for the reviewer is simply in testing a new product to the extreme, and he makes no comment as to how the digital glitch sounds, even as to its possible grating quality. However, as we will see in the next section, Ranada's methods for testing the system are extremely close to the technique used by Yasunao Tone to produce his indeterminate compositions from "wounded" CDs. The major difference between them is the musician's desire to aestheticize the errors that occur at the breakdown of the technology.

2 Yasunao Tone's Wounded Compact Discs: From Improvisation and Indeterminate Composition to Glitching CDs

Yasunao Tone is a Japanese experimental musician with a substantial history of practice.[13] Although his use of indeterminate techniques dates back to the early 1960s, he is also at the forefront of the current interest in glitches, cracks, and unstable systems for sound production, all of which use a measure of indeterminacy and chance. Tone was at the center of the Tokyo Fluxus movement in the 1960s and had close links, originally through Group Ongaku, with Fluxus artists including Ichiyanagi Toshi, Kosugi Takehisa, Nam June Paik, and Yoko Ono, as well as composers such as John Cage and David Tudor. Tone's compositions are harsh in their intensity and volume but are also compelling in their radical use of indeterminacy to generate unexpected outcomes. He came across the CD glitch in 1984, and has since included it as a dominant and extremely productive element in his composition and performance work.

Historical Context

Japan—especially Tokyo—was a hive of artistic activity in the 1960s, with various branches of contemporary and avant-garde practice connected with movements in Europe and America. As Alexandra Munroe writes:

> The correspondence of thought and attitude among Euro-American and Japanese artists in the 1960s was no coincidence. The activities of both groups represented an interest in early twentieth-century anarcho-cultural sensibilities, specifically Dada. In Japan as elsewhere, this legacy taught outrage against the staid idealism of bourgeois culture and prompted the defiant pursuit of an aesthetic of negation of art: negation of art as an illusion rather than presentation of reality: and negation of the boundaries between object and action, word and image, art and life.[14]

The work of composer/performer Ichiyanagi Toshi was a significant influence on Tone's development, as well as other Japanese-based Fluxus-related artists. Ichiyanagi was well versed in the various theories that were to constitute Fluxus, especially Cagean theory. Ichiyanagi had studied at the Juilliard School in New York between 1954 and 1957 and attended John Cage's classes at the New York School for Social Research in 1959. At this time, Ichiyanagi and his wife Yoko Ono were regulars in Cage's circle.[15] Ichiyanagi returned to Tokyo in the summer of 1961, bringing with him current ideas from America, and encouraging Tone and others producing similar work at the time. The year of his return is considered a watershed for experimental music in Japan, and it was witnessed as an "emergence of an overflowing experimental spirit," according to Ichiyanagi.[16]

As a movement, Fluxus fully embraced the Tokyo art scene, and a number of Japanese artists had close links to Fluxus, including Genpei Akasegawa, Ay-O, Ichiyanagi Toshi, Kosugi Takehisa, Shigeko Kubota, Nakanishi Natsuyuki, Yoko Ono, Takako Saitō, Shiomi Mieko, Takamatsu Jiro, Yasunao Tone, and Wada Yoshimasa.[17] Tone's own practice as a composer/performer initially developed during the productive period of the international Fluxus movement.

Born March 31, 1935, Yasunao Tone studied at the national Chiba University in Japan, but he did not study music. He recalls:

> I simply sat some classes of musicology in the Tokyo Art College. I started playing alto saxophone first, and other instruments that belonged to the ethnomusicology department. I studied literature and wrote a graduate thesis on Dada and Surrealism in my college.[18]

Tone was a member of Group Ongaku[19] along with Kosugi Takehisa, Mizuno Shūkō, and Shiomi Mieko.[20] The group was founded in August 1960 as an improvisational music unit, although they had been playing together before this date. Mizuno was a classmate of Tone's from Chiba University and introduced Tone to Kosugi Takehisa after they had graduated. Kosugi and Mizuno had started a duo and asked Tone to join, even though he had no knowledge of how to play any instruments. This was not a problem for Kosugi, who sold Tone his alto sax. During this time, while the Tokyo National University of Fine Arts and Music was a conservative university, a small group of students involved in the art and ethnomusicology departments is of particular interest. The composition and ethnomusicology group met regularly for improvisational sessions. Tone describes these sessions:

> We played together in the ethnomusicology department studio. We found many ethnic music instruments on top of

> western musical instruments. Before long we realized we were playing any instruments we could touch. That extended to any objects that made sound. We assumed what we were doing was kind of automatism and our way of making sound was quite similar to Abstract Expressionists and Signifiance de L'informer painters. We were recognized as vanguards of contemporary music by musical journalism and we had very good responses from visual artists.[21]

The first group concert by Group Ongaku was entitled "Improvisation and Objet Sonore" and was held at Sōgetsu Hall, in Tokyo, on September 15, 1961.[22] The performance was the group's first formal outing, though they had previously played impromptu concerts at such places as the Tokyo Pier, in a car while driving on the freeway, and often at friends' houses. These performances can be heard as some of the first instances of what we now call free improvisation.

The debut was to be Group Ongaku's first and only public performance as a collective, owing to a disagreement about presentation. The performance was described by Tone as "a rumble of sounds, leftover sounds, actions, water, bells, glasses, blowing."[23]

Throughout the 1960s, Tone was involved in a range of activities centered in avant-garde practice. His own intermedia practice encompassed composition, performance, event-based work, theater, dance, and criticism. Tone arrived in New York in June 1972 and continues to be active in the arts scene. He has composed a number of pieces utilizing indeterminate techniques for the Merce Cunningham Dance Company, including *Geography and Music* (1979), which was performed with the dance work *Roadrunner* (which premiered July 19–21, 1979, Duke University, Durham).

Tone's Indeterminate Compositions and the Everyday

Tone's early work of the 1960s was heavily involved in indeterminate composition, the outcomes of which resemble the "event" scores being written in the United States only slightly earlier (although they are not as poetic and are more closely aligned with music). The artists most closely aligned with the event score were La Monte Young, George Brecht, and Yoko Ono. Their texts can be read in a number of ways: as music scores, visual art, poetic texts, performance instructions, or proposals for action.[24] The scores, in typical Fluxus manner, question distinctions between the everyday and art. For example, the most simple event score, George Brecht's *Word Event* (1961), consists of the instruction "Exit." Just what could merit a performance of this work is up for interpretation—it could be any exiting action made at any time. The boundaries of art and life are here blurred.

Tone's work since the 1960s has focused on exploring various methods of composition that introduce random events and indeterminate compositional techniques into live performance. His score for *Anagram for Strings* (1962), for example, directs performers to draw a line that intersects with various circles; on the intersection, the performer is asked to play different downward glissandos. *Geodesy for Piano* (1962) asks performers to place an acetate sheet with lines over a topographical map and work out various heights, angles, and positions from which to drop objects from a ladder onto the strings of a piano;[25] in *Molecular Music* (1983), the performers fix photosensitive detectors to a wall and project characters and photographs onto the wall, the detectors being in turn connected to oscillators; *Paramedia Music* (1990) directs performers to record the phone numbers of various people in the audience of a concert, then call them and use the voice messages on their answering machines as the audio source. These works point to Tone's interest in the boundaries of systems and their various outcomes, and how these can be used to create compositions.

Tone has said that Fluxus

> didn't try to make art, but to abolish the boundary between art and life. So we made everything no-art-art. Artists made calculations, or slept for a while . . . washing your face, everything could be art.[26]

It is through this interest in the everyday that we will see a connection to the use of everyday technology and tactics as outlined by Michel de Certeau in chapter 4.

Tone's interest in the sounds of the everyday can also be heard in the recording of Group Ongaku's piece *Automatism* (1960). The work was recorded at Mizuno's house and utilizes musical instruments—piano, pedal organ, alto saxophone—as well as everyday objects—a vacuum cleaner, an oil drum, a radio, dolls, and a set of dishes. The group had their ears open to the noises of the everyday and found the materials of sound in whatever could produce it.[27]

Yasunao Tone's "Wounded CDs"

In 1984, Tone came across a method that he thought could be used to cause discrepancy in the playback of the then very new CD technology. While working on a composition, Tone read a book entitled *Science Seminar for the Familiar*,[28] where a chapter on digital recording included this paragraph:

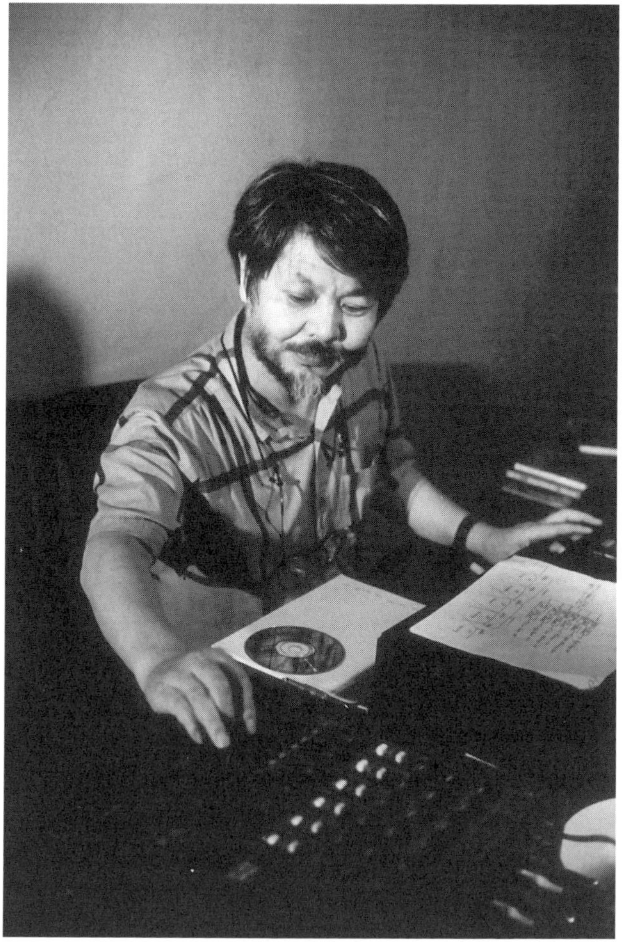

3.2

Yasunao Tone, performing *Music for 2 CD players*, DIA Art Foundation, New York, June 1987. Photo: Paula Court.

> Digital recording is a wonderful audio technique, since it has almost no noise and produces sound very faithful to the original. However, when it misreads 1 with 0, it makes very strange sounds due to the binary code becoming a totally different numerical value.[29]

This passage exemplified the emphasis that was placed on the lack of noise in digital recording technology. It hinted, however, that the new technology could be used to make noise by overriding the error correction system. Tone recounts:

> I called my audiophile friend who owned a Swiss-made CD player and asked about it. It was a simpler method than I suspected. I bought a copy of Debussy's *Preludes* and brought it to my friend's place. By his engineer friend's suggestion, we simply made many pinholes on bits of Scotch tape and stuck it on the bottom of a CD. I had many trials and errors. I was pleased with the result because the CD player behaved frantically and out of control. That was a perfect device for performance.[30]

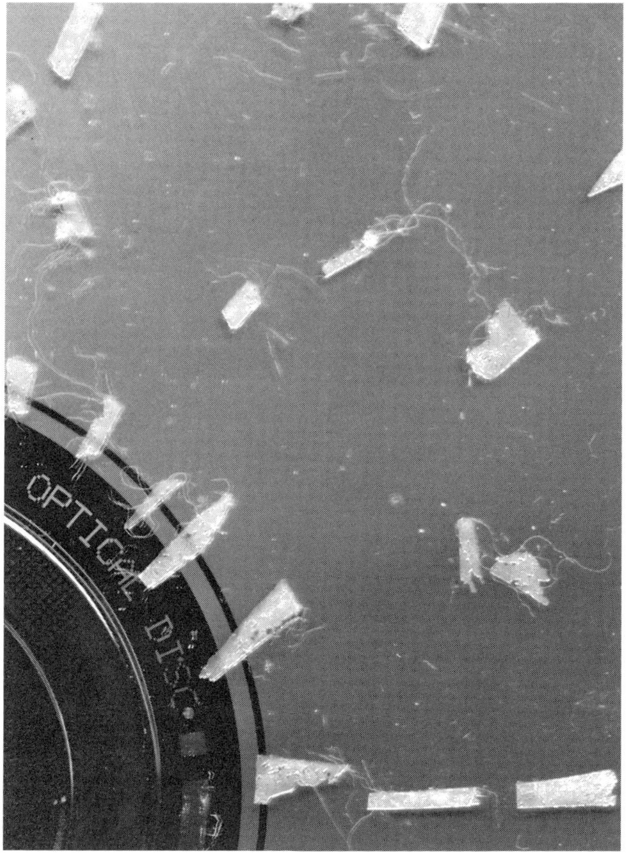

3.3
Yasunao Tone, prepared CD.
Photo: Gary McCraw.

The result changed the pitch, timbre, rhythm, and speed of the original piece. The CDs also produced a "stuttering" that was different each time the disc was played. The stuttering CD extended the possibilities of performance in a number of ways. The sounds produced by the manipulated discs are never quite the same. This means that in performance, the performer has no clear idea of what is going to happen, making such things as scoring very difficult. Tone commented, "Playing Prepared CDs according to the score was like advancing in a maze where ambush was everywhere, and that made the performance situation all the more interesting."[31] Tone knocks and jolts the CD tray during performance, gently nudging it to continue reading the prepared CD; just where it will stop and on which sound it will stutter cannot be known from one playing to the next. At the first performance of *Music for 2 CD Players* (1985) on March 16, 1986, it is quite conceivable that many in the audience had never actually seen a CD player, and yet, here was a composer pushing it to the point of breakdown, causing it to madly and loudly stutter and skip.

Here lies the indeterminacy of the piece. It is not completely random, as it is still working within a system designed by Tone, but the outcome is not fixed and the precise manner in which the CD player will handle the wounded CD cannot be guessed. The technique has been extremely productive for Tone, who has used it from 1985 through to the present day. He explains:

> A new technology, a new medium appears, and the artist usually enlarges the use of the technology. . . . Deviates. . . . The manufacturers always force us to use a product their way . . . however people occasionally find a way to deviate from the original purpose of the medium and develop a totally new field.[32]

Thus, one year after CD technology was released in the United States, Tone, much like Ranada, had devised a way to override the error-correction system, and the mechanism designed to allow seamless playback was forced to glitch. The idea of a playback technology that could play pure clean audio was displaced by Tone's noisy, glitching CDs. The silence of digital audio was made to produce noise; the purity of the new medium was damaged, and this technical imperfection was exploited as a performance tool to create an outburst of abstract audio.

When Tone was first approached to release a CD of his music, he believed that his pieces were not appropriate for release. Because of his disavowal of the representation of performance and his belief in the direct connection between performance and improvisation, Tone had little interest in releasing recordings of these performances. Tone is well aware of Cage's approach to the recording medium and his issues surrounding both live and recorded performance, as well as those raised by the reproductive process of recording and its static repetition. Like Cage, he is more interested in the live performance, or the one-off, than in the recording of the event. As Cage wrote:

> A performance of a composition which is indeterminate of its performance is necessarily unique. It cannot be repeated. When performed for a second time, the outcome is other than it was. . . . A recording of such a work has no more value than a postcard; it provides a knowledge of something that happened, whereas the action was a non-knowledge of something that has not yet happened.[33]

In response to this dilemma Tone made a piece that could not be played live, and was totally predicated on the CD release for its realization. *Musica Iconologos* (Lovely Music, 1993) is Tone's first released recording and is a composition comprised entirely of computer-generated snippets of noise. *Musica Iconologos* is described by Craig Kendall, in the liner notes to the release, as "one of the most extreme and original applications of the current digital recording medium."[34] For the piece, Tone turned the characters of the Chinese poems "Jiao Liao Fruits" and "Solar Eclipse in October" into sonic analogues.[35] The piece was created from scanned images of Chinese characters and photographs, which Tone believed to represented the script. One hundred eighty-seven scans were run through an "optical music recognition" program, and the sound files produced were processed digitally, with no personal judgment used to exclude or repeat any of the audio. The piece moves through the characters in brief bursts of digital audio, creating an extreme noise composition. Tone states that this CD did not exist as music until it was mastered, so that the production of the CD was part of the work's process:

> The result was noise in all senses . . . when you play that CD, what you receive is not images as message, but sound which is simply an excess . . . [one of] the French word[s] for noise, "parasite," indicates [that] it is parasitic on a host—that is, message. But, in this case there is no host, only [a] parasite on the CD. Therefore this CD is pure noise.[36]

Musica Iconologos created a problem for Tone, as the piece existed solely in the digital recording process and could not be played live. Tone, although mostly interested in live performance practice, had created a major work that could not be played live in a way that interested him, as the recording was not open to any chance occurrences or creative possibilities in its playback. His answer to this problem lay in his use of the wounded CD: he took copies of the *Musica Iconologos* CD and prepared them with scotch tape, performing the piece with CD players. The work was radically remixed as it stuttered, skipped, and jammed throughout the performance. The piece now met the conditions that interested Tone in performance, those of indeterminacy and chance. Tone had set up the conditions for the piece, that is, he generated and prepared the CDs, and then let them loose in the playback equipment. In this sense, the piece is indeterminate as Tone has final control not of the outcome of the performance, only of its limits. He has, however, released a recording of a performance of the piece, entitled *Solo for Wounded CD* (Tzadik, 1995).[37] This is a "remix" of his original computer-generated noise through the stuttering of the damaged and modified CD. In this case, Tone in a somewhat contradictory manner sees the release as simply a documentation of a performance.[38]

Tone's technique of remixing and reworking his releases was also used in a 2000 work entitled *Wounded Man'yo 2/2000*. The work, which won an honorary mention at the 2000 Prix Ars Electronica, is based on his CD-R *Musica Simulacra*.[39] *Musica Simulacra* is made in similar fashion to *Musica Iconologos*. The software Projector was used to scan images formed by Tone to relate to the text *Man'yo-shu* (this text, a collection of some 4,500 poems compiled in the eighth century, is the oldest anthology of poetry from Japan). For *Wounded Man'yo 2/2000*, Tone recorded a number of CD-Rs from the original CD-R of *Musica Simulacra*. He then "wounded" the CD-R copies and used the prepared discs in performance.[40] At this point, then, the final recordings had gone through many levels of indeterminacy.

Tone has been involved in contemporary compositional practice since the early 1960s. His methods have much in common with contemporary digital music practices and his work is appreciated within a contemporary new media context, as evidenced by his Golden Nica award. Tone's practice thus traverses a number of important periods in experimental sound practices, from the latter half of the twentieth century and into the twenty first. Tone connects the John Cage–influenced Fluxus movement and post-Cage aesthetics to contemporary practice more directly than other Fluxus sound artists. It is through his interest in indeterminate composition and the digital error in CD playback that we can most clearly see this connection. His interest in noise is now performed at an optimum level through large PAs at international festivals such as Sonar in Barcelona. His use of extreme volume, and the indeterminate method for performance he developed, connects closely with many postdigital practitioners of glitch, as discussed in the introduction. One development of glitch was that many of the indeterminate features of the style were lost as the noisy sounds were sampled and sequenced; any risk or chance was lost in a perpetual loop. Tone, however, has never stopped using indeterminate techniques, and his development of the "wounded CD" technique has been an extremely productive approach in his later compositions.

In 2004, Tone release a duo recording with Florian Hecker entitled *Palimsest* (Mego). Hecker is a German computer musician who works with chaotic software "patches" for performance and produces extremely noisy and seemingly random digital sound. The combination of Tone's contemporary practices with Hecker's work further demonstrates the connections Tone has made with the contemporary digital music scene.

3 The Unmuted CD Players of Nicolas Collins

Like Tone, Nicolas Collins's approach to technology and his use of CD players for performance shows an interest in indeterminacy, chance, and the accident. Collins differs from Tone, however, in that his approach involves tampering with the CD player's internal mechanism rather than corrupting the disc. Collins has been using circuits and modified electronic equipment for the creation of music since the 1970s;[41] his work is linked to a generation of composers, including Alvin Lucier, Gordon Mumma, and David Tudor, who experimented with electronics in such a way as to shift the musical score from paper to the circuit board. Thus the approaches of Tone and Collins represent two major strands of twentieth-century experimental music and sound, which join here at the point of the crack and break.

In 1988, Collins opened up a CD player and tinkered with its circuitry. In a 1995 interview Collins recalled that he "looked at the CD player as a challenge. I prefer vinyl in almost every aspect, but 5 years ago it looked as though the CD would take over. I took it upon myself to try to corrupt this 'perfect' medium."[42] His work with CD players came about as part of a wider interest in electronics as a type of score, the recycling of sound, and "the benevolent catastrophe."[43]

Nicolas Collins was born in New York in 1954 and studied with Alvin Lucier at Wesleyan University; he also worked with David Tudor at this time. Tudor was highly influential on Collins's subsequent interest in electronics and sound systems. Collins sees Tudor as "the inspiration behind a whole kind of movement in America, of composers who looked inside technology to generate a piece of music."[44] Collins believes there are two distinct approaches to the use of technology when composing a work. The first is to take a preexisting piece of technology and look inside it. He compares this to Michelangelo looking for a sculpture inside a particular piece of stone rather than having an image and forcing that onto the stone. The second method is to design a circuit from the bottom up. Collins explains, "designing a circuit was like composing a piece, the piece and the circuit were the same thing. The circuit was the score, the circuit was maybe your performer."[45] The circuit is thus the actual composition and not merely an instrument; the circuit diagram becomes a self-performing graphic score. Collins states:

> I think that a lot of my music has had to do with the implications present in a piece of technology, even very common circuits, consumer electronics: I take a CD

> player, I modify a radio or a walkman, trying to work at a very low technological level and then customize it a little bit, cannibalize it. But I've also built machines up from scratch, and have also done a fair amount of work with computer programs. It's all part of the same stream flowing from the notion of compositions or implications present in technology.[46]

Collins points to a number of key issues in the above quote. The compositions or performance possibilities are understood to be present in the technology itself, and his role as a composer is to draw them out. This is carried out by manipulating the already designed and produced technology, in the process cracking it by disabling certain parts of the mechanism, or breaking it so that the composition can be generated. Once the system has been cracked Collins inserts information, the data on a CD in this case, and allows the technology to direct the data through the rerouted system.

During his study of composition, Collins witnessed the "trickle-down" of affordable electronic technology. Electronics were becoming cheaper and as such were available to a wider audience to use and experiment with, as he notes:

> Before 1970 few composers were known for making their own electronic instruments (in America, notably David Tudor, Gordon Mumma, and David Behrman), but [the] proliferation of electronic components and information in the early 1970s caused the widespread development of idiosyncratic "home-made" electronic musical instruments.[47]

A key word here is "idiosyncratic"; electronics manufacturers create their product with a certain audience and use in mind. They include only what they see as useful or necessary. However, a musician or artist who creates his or her own electronics, or modifies already existing ones, may have uses and outcomes in mind not thought of by the manufacturers. The process itself, though time-consuming, leads to "instruments that had a distinctly personal character, that we could afford, and that often were much more powerful (if more specialized) than anything commercially available."[48] This observation has a clear link to the development of turntable practices that would simply not have occurred without a surplus of vinyl, and without a devalued and widely accessible playback technology.

Nicolas Collins and the Manipulated CD Player

Collins corrupted the compact disc medium by manipulating the internal mechanism of the CD player itself. Without any real knowledge of the workings of the technology, Collins assumed correctly that the CD player's laser never left the disc's surface, and that it read not only audio information but "hidden" information such as error-detection and information-coverage data, as well as data defining track locations, lengths, and so on. Locating the player's control chip, Collins came across a "mute" pin, which he removed, resulting in a constantly chattering playback. Detailing this process, Collins explains:

> With this pin removed, the CD player never shuts up, and one can hear the sound as the laser "scratches" (a magnificent, cartoonish ripping noise) or "pauses" (fast looping rhythms, possessed of a peculiar stutter and swing). Further modifications to the control system, done in collaboration with engineer Sukandar Kartadinata, opened up more extreme aberrations of digital misbehavior.[49]

With this pin no longer in place, the CD player reads all the information as audio. It is never quiet as it moves across the disc, reading and playing information such as the track numbers and track lengths as audio. Even while it is in pause mode it is still reading the information that keeps it in place. It would also play all the error-correction data and the noise usually muted by the correction software when covering for lost data.

Collins at the time saw interesting audio possibilities in the irregularity of the system and the rhythms that were produced, "because unlike on a record they are not regular, they hiccup or stutter."[50] His interest in the sound of the modified CD player comes from an interest in hi-fi sound contrasting with the "snap crackle pop of the digital click," and the "automatic rhythmic variation of the skipping occurrence."[51] As with Yasunao Tone's stuttering CDs, it is never certain just how they will cope with errors—perhaps looping, jumping, skipping, stuttering, or hanging for a moment before veering off.[52]

What makes the skipping and stuttering sounds of Collins's work different from those Tone is that Collins has actually modified the CD player itself. The stuttering is not coming from corrupting the disc but from tampering with the CD player. This aligns with Collins's connection with earlier electronic experimentalists from the American avant-garde, and the notion of circuit board as score. Collins is

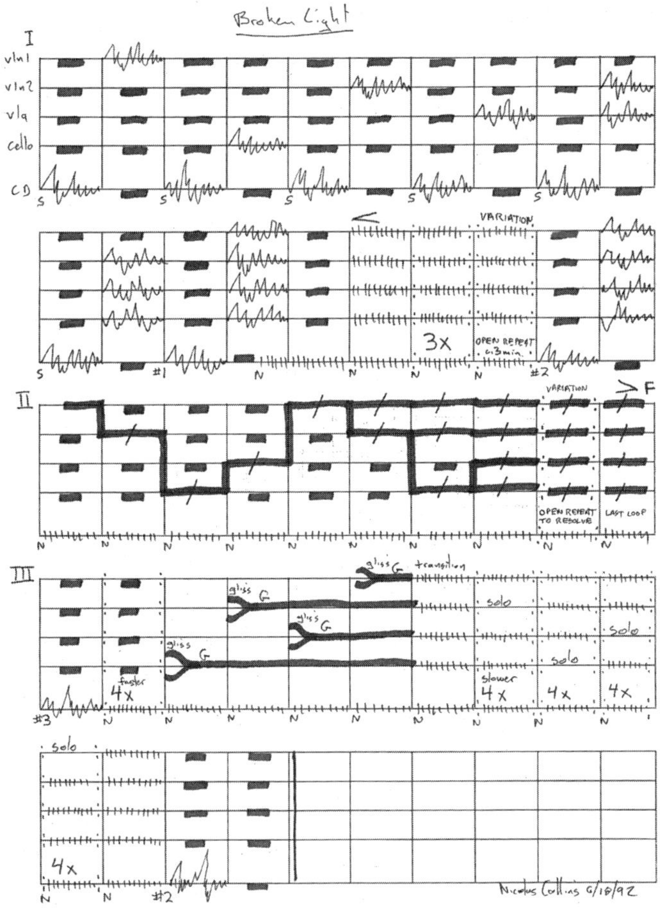

3.4
Nicolas Collins, excerpt of *Broken Light* (1991–1992), score of the first composition for skipping CD and string quartet. © Nicolas Collins.

not actually breaking the system; rather, through the technique of the crack he can be heard to be extending and manipulating it. We will see in the next section that Oval, and its brand of skipping and glitching CDs, comes from yet another historical lineage, that of indie pop.

4 Oval: Pop into Art Music

> Oval's "music" offers an uncanny, seductive beauty of treacherous surfaces and labyrinthine recesses.
>
> —Simon Reynolds[53]

> The ability to create new content by sabotaging the linear playback of data has proven to be a successful interface for Popp, who has double-clicked an entire genre of electronic music using it.
>
> —Kim Cascone[54]

Oval is the quintessential CD glitch band. In the 1990s Oval gained a reputation for its signature use of the CD skip, its brand of glitch instantly recognizable for its clean, electronic, and almost ambient sound. The group's music was taken up by the Tokyo fashion show catwalk and was the soundtrack to an international advertising campaign for Georgio Armani cologne. The music Oval produced has become strongly linked to the musical soundscape of the mid-to-late 1990s.

This section will look at just how Oval "double-clicked an entire genre of electronic music." Oval is heard as the forebears to the use of cracked media in the mid- to late 1990s digital audio scene. The group's audio has sparked interest in the various methods by which the breakdown of digital and analog audio can be used for musical experimentation. The importance is not that they were the first to use the skip—as we have seen throughout this section, they were not—but that they introduced the idea to a new audience, an audience that was computer literate and looking for new ways to produce sound in the digital environment.

Oval is situated a long way from the compositions and performance of Tone and Collins. It is not part of the musical avant-garde but instead hails from the indie-pop and rock movement of the late 1980s. *Systemisch*, the group's most influential and successful album, can be clearly heard as skewed pop; the pieces retain the basic structures of pop, with melodic motifs and repeated phrases formed into verse- and chorus-type structures. Perhaps it is because of this accessibility that Oval became so popular in a subculture intent on new sounds and approaches to music, while still needing something familiar to bring listeners in.

Oval's Prehistory

Oval's original core comprised Frank Metzger, Sebastian Oschatz, and Markus Popp. Oschatz and Popp formed a short-lived band called Die Lichter (The Lights) with Silvana Battisti in 1989. The band was influenced by British indie pop and recorded all its improvised rehearsals, a later trademark of early Oval recordings; the group collected short sketches that would later be used as the basis of longer, more involved tracks. Whereas the music of Die Lichter was guitar-based pop, one unreleased track, "Lavita," is of particular interest; it prefigures Oval with a continuous oscillating tone that plays in the mix and becomes dominant by the end of the track.

Metzger met Oschatz and Popp in Darmstadt in the summer of 1989 and started playing with them in 1990. At that time eight to ten people would regularly meet at Popp's house to take part in improvised sessions. Popp played guitar and bass, Oschatz played keyboards and percussion, and Metzger sang. Popp brought along an Atari computer and a sampler, and sequences were left playing during these meetings. It was from these improvised jams that Oval developed.[55]

Another work produced by the trio that directly prefigured Oval was the soundtrack to a short film made by Metzger entitled *Mrochen*. It has oscillating tones playing throughout the duration of the piece, and is full of space with a stripped-back feel. This open style may well be the result of the freer nature of a film soundtrack, in that it allows space to create music not based solely in pop structure. Although Oval would not head in this direction for another three years, this work illustrates a seed of interest in an abstract sound space.

Preceding Oval's first full-length release *Wohnton* (Ata Tak, 1993), three limited-edition cassette tapes were produced entitled *Azimut*, *Oktober*, and *Kolor*. These cassettes display elements of pure guitar pop with Popp's clean bass sound and Metzger's improvised and almost comic vocal style. Other tracks use samples and sequences with less guitar work, if any at all. In addition, "Shh" from *Oktober* clearly uses a CD played on fast-forward through a voice sequence, which stutters throughout.

According to Metzger, the first piece to use the glitch was "Neopren," which appears on the *Oktober* tape. This initial CD stutter effect happened by accident. Metzger had borrowed a jazz CD from a public library that was in terrible condition. The CD stuttered and "created interesting 'glitches,' mostly because it wasn't simply hanging as it would usually do, the 'stutter' was moving, so the pattern was constantly changing randomly. Later we tried to reproduce this effect."[56] The glitch appears in the last thirty seconds of the track, forming a heavily affected sound to end the piece.

The sound of the glitch became Oval's signature sound. The group sampled the audio produced when the CD player was on fast-forward, looping and dropping the pitch by one or two octaves. Oval used Metzger's sister's CD player, as it had the best sound of any they had access to.[57] The CD player was a very cheap "no name" machine that Metzger still uses. The idea of better-sounding glitches is of particular interest here, as the sound of a CD player in fast-forward mode is heard to differ from machine to machine, and from this particular player a very specific aesthetic was developed. If the sound of the glitch is in fact noise, then we have here an aesthetic based first in the usually unwanted sound, but we also have a hierarchy within that area—a preference for one noise over another. Oval was listening to, and discriminating between, the different noises that CD players produced when the technology was pushed, manipulated, and cracked.

Subsequent to the discovery of the skipping and stuttering library CDs, Oval began to prepare CDs with nonpermanent marker pens and small strips of adhesive tape to force the CD to glitch. This technique is reminiscent of Yasunao Tone's "wounded CDs," although importantly Oval is not using CDs that the group itself produced. Nor is it important exactly what CD was used; only the actual sound of the glitching and stuttering CDs was important to the group. Oschatz explains that:

> Later on Markus literally rented every CD from a CD-rental service around the corner; played everything in fast forward, painted the CDs, took 1:1 samples and brought them back. In practice we had no relation to the material and seldom really knew where the material originated when we used it.[58]

Like Tone and Collins, Oval is here deliberately producing noise from the technology, and like Tone the group did not know how the CD player would react to the prepared CDs, only that the preparations would overload the technology's error correction. The prepared CDs may skip or stutter, they may play differently each time they are placed in the CD player, or they may not play at all. The nature of the skip itself, however, actually gave Oval trouble, confronting the group with the need to determine new methods of improvising with the stuttering rhythms produced by the glitching CDs. The band was excited by the complex rhythms and the nature of the clicks and glitches, but the results were nearly impossible to improvise to. Oschatz recalls:

> This was mostly because of the complex and full sound (because the samples were taken from fully mastered songs, which already utilized all frequencies). We tried this extensively but they really didn't leave enough space for adding other instruments.[59]

This is an important observation as at the time Oval was a band experimenting with traditional pop arrangements and the addition of electronic elements, including the newfound skipping CDs. Just how the band was to use these new elements became a key issue. When faced with the full and rich sound that made up the basis for the glitches, it did not seem possible to add more instrumentation. The computer application Cubase, a MIDI music sequencer and digital audio editing application, was introduced as a sequencing tool for the newly acquired skipping CD samples, and was developed further in Oval's third tape *Kolor* (1993), which later became their first official release *Wohnton*.

Wohnton: Glitch-Pop

Wohnton, released in 1993 on the German label Ata Tak, was Oval's first full-length album. The release maintains a close connection to traditional indie pop and includes lyrics and vocals by Metzger. It was, however, the developing use of the skipping and stuttering CD that caught the attention of reviewers at the time. Jörg Heiser, in the wonderfully entitled review, "dechingdechingdeching," hears various themes in the music of Oval that would follow the group in subsequent reviews and articles. Heiser picks up and focuses on the CD skips as a key factor in the group's sound:

> [Oval has] shown that after the CD skip . . . a new beauty of mistake developed: Now the noise was emancipated from the original signal. . . . Oval have used this to their advantage and have drawn fine lines onto their CDs with felt pens in order to continuously record new Dschingdschingdsching and CD searching sounds, to put them into the computer and in this way to have access to further processing in the

> midi-system (the electronic combination of PC, sampler, keyboard, sequencer etc.).[60]

Heiser observes that the CD skip still has a grasp on the memory of the original sounds found on the CD. The idea of the memory of the affected sound has a direct link to the memory of the manipulated and cracked vinyl record. Although these sounds have been refigured and manipulated, they still have a connection through their very data to the original recording.

In an astute reading of *Wohnton*, Heiser argues that Oval produces an "ensemble of noises set by a computer which is not immediately music." He explains that it is possible to choose "nice passages" and layer them over each other, and that the looped and repeated nature of this process will create music out of what is otherwise simply noise. He also argues that this is neither an attempt at liberation nor a destructive attitude. Oval, as previously noted, is not destroying anything in its processes, as the marking pen used to cause the glitch is cleaned off after the results are sampled. Oval, according to Heiser, has only temporarily "destroyed" the CD's clean and stable playability by covering sections of information with marking pens and by sampling the results, and in doing this have created "unintended beauty from intended destruction."[61]

How is this observation of beauty in the glitch possible when we know just how harsh, grating and irritating the sound of a skipping CD can be? It is through the sampling, looping, and layering that the beauty effect is achieved. The unknown of the accident is layered into the known of the pop format, the choruses and verses underpinned with the accidental sounds of stressed digital audio. Through repetition we begin to understand the mistake and aestheticize it. The notion of the crack and break as transgressive is undone in the work of Oval.

In another review of *Wohnton*, Sascha Kölsch writes about the group's use of MIDI and Cubase technologies, suggesting that Oval is unhappy with the applications because they try to dictate what can and cannot be produced with them.[62] This reading is in opposition to Heiser; Kölsch is arguing that because Oval is not using the systems in a straightforward fashion, the group is therefore unhappy with the system's original intended uses and methods for use. This notion of Oval as unhappy with the conditions of the electronic studio will be discussed later in the section and will also emerge again in Simon Reynolds's article on the scene promoted by the German record label Mille Plateaux.

Pop into Art

In 1993, Oval created an installation entitled *Wohnton*, which used the extended reworking of a number of tracks from the band's *Wohnton* album for its soundtrack. The exhibition was first shown in November 1993 in Arnhem, the Netherlands. It consisted of 128 miniature speakers within eight modules, which in turn were connected to an eight-track playback system, allowing the sounds to move around the listeners as they themselves moved around the exhibition space.[63] *Wohnton* was exhibited in a number of galleries, as well as dance parties, between 1993 and 1996.[64]

The installation is the first time that Oval produced work much longer than standard pop duration. The tracks used were extended versions, with vocals removed, of the earlier pieces "Alles in Gedanken," "Hallo Draussen," "Kardamom," and "Hallo Draussen" played backwards. These pieces were made into fifteen-minute-long tracks consisting of three- to five-minute loops. This sound installation was the first sign that Oval was progressing, as the group shifted its focus from improvised pop to experimental art music.

A possible reason for this very rapid transition is the *Wohnton* installation itself. The work provided an opportunity to experiment fully with a form different from that of an album. The very nature of

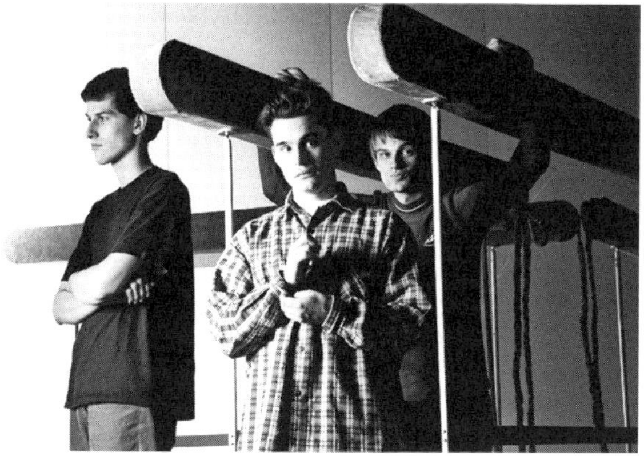

3.5

Oval. Left to right: Frank Metzger, Markus Popp, Sebastian Oschatz, in front of *Wohnton*, Hamburg, 1994.

the installation offers a longer and more abstract format. Members of the gallery audience would have been able to enter the work at any point within that twenty-minute cycle and understand what it was that they were experiencing. This experience propelled Oval to remove the vocals altogether and triggered a shift toward a much more abstract music in a very short period of time.

Systemisch and 94diskont

Oval's second album *Systemisch* (Mille Plateaux, 1994) was the first release to gain wide attention: it was reviewed in the mainstream media including the *New York Times*, *NME*, *Melody Maker*, and *Billboard*.[65] The release combined experimentation with the familiarity of pop, and it indeed became very popular—the CD is still in print at the time of writing. *Systemisch* completely dispenses with vocals as well as the use of actual instruments, and unlike the *Wohnton* installation it does not have any extended, long pieces; instead it focuses on tracks under ten minutes long to develop its themes.[66] This release clearly centers on Oval's use of the sequenced skipping CD.

Up until *Systemisch* the sound of the distressed CD had only been heard by most as annoying. This album made the sound not only palatable but also a thing of beauty. The unplayable CD, only good for the scrap heap, had become the content of musical production. The traumatic sound of the skip had exposed the myth of the CD format being indestructible, and Oval had aestheticized that sound, made it listenable. Once again this is very different from the audio produced by Yasunao Tone, whose wounded CDs still have the ability to cause intense irritation, even for an audience of experimental music cognoscenti. The pieces on *Systemisch* seemingly stutter, as if caught in the looped skipping and repeating rhythm of the CD, while programmed bass plays under the mid- and high-range skips and glitches, at times forming a regular clapping sound. The complexities of the sounds caught and sequenced by Oval are what make them so appealing; they are full of texture, the paths of the tracks compelling. Oval has taken the CD's failure and made it musically accessible.

As such, Oval's audio represents the planned use of glitches. No longer is the glitch an unexpected or even a chance occurrence, as it is sampled and sequenced. The glitch is recorded, so that it can be repeated an infinite number of times on cue. We might picture the group standing around a CD player, listening to their CDs skipping away until at some point it makes just the right skipping sound, repeats, and clicks just so. When this happens the sounds emitted from the confused CD player are recorded, sampled to be used over and over in the repeating and looping tracks, and overlaid with other prerecorded skipping CD sounds. There is no need to score the prepared discs as they are never played live: an Oval performance uses sequencing and mixing software on computers, the actual skip a distant memory.

The track titles on *Systemisch* offer further insight into Oval's work. The titles do not describe the atmosphere or mood of the tracks but rather the working conditions under which the tracks were produced. They allude to the context of desktop music and the issues confronting Oval at the time.[67] The titles came chiefly from Popp and make various references, both obvious and obtuse. Titles such as "Aero Deck" came tangentially from the original working title "Aero Deco," which is a spray used to decorate windows so that they look frozen. "The Politics of Digital Audio" makes reference to a 1980s pop music track by Re-flex entitled "The Politics of Dancing," as well as literally suggesting that there *is* a politics of digital audio. More telling is "Catchy DAAD," which is an abbreviation of either "Digital/Analog Analog/Digital" or "Deutscher Akademischer Austauschdienst" (German Academic Exchange Service). This refers to Oval's method of sound production. The group records digital skipping to analog tape and then digitally samples the tape. In addition the title also makes reference to the pop music ideal of a catchy track, or a catchy chorus line. Though the track is not catchy in the traditional sense, it is quite possible to imagine getting the track stuck in your head and spending the rest of the day singing the main theme.

As these titles, coined by Popp, suggest, Oval had a reflexive awareness of their own tools and methods. At the time, the group's members saw themselves as not so much the makers of music but rather as organizers of database systems and strategists for information management. Popp tells us in an interview with A. C. Lee that

> Oval are less authors than an interface, which searches to coordinate itself musically with regards to the utilized system. Fundamentally, we deal less with music than with the technical implementation of music, less with concepts and formal differences than with the concrete shaping of its content.[68]

In Popp's statement above, he describes his goal to work with systems, with user interfaces, and audio files. This approach and its ramifications have been observed by Lev Manovich in *The Language of New Media.* Manovich argues:

> Because new media is created on computers, distributed via computers, and stored and archived on computers, the logic of a computer can be expected to significantly influence the traditional cultural logic of media; that is, we may expect that the computer layer will affect the cultural layer. The ways in which the computer models the world, represents data, and allows us to operate it; the key options behind all computer programs (such as search, match, sort, and filter); the conventions of HCI [human–computer interface] . . . influence the cultural layer of new media, its organization, its emerging genres, its contents.[69]

Manovich discusses the implications of computer storage systems and databases (the idea that the language of the computer and its functions are actually influencing the cultural layer of society). In a cyclic relationship, the interface metaphor of files and desktops resembles older paper storage systems, influencing computer-aided production in both the workplace and the social sphere. In the same way, Popp is interested in the graphical user interface and the effect of new media metaphors on his audio production. This was to become a key concern for Popp, who took the Oval project as far as the creation of an installation entitled *Ovalprocess* aimed at allowing users to create their own Oval tracks.[70]

In 1995 Oval released a companion album to *Systemisch* entitled *94diskont* (Mille Plateaux, 1995). These two albums were meant to display the full scope of the Oval project during 1994.[71] With *94diskont* Oval's pop music faded even further; the release had more in common with the *Wohnton* installation than with the more pop-based *Systemisch* album.

The main track on the album, "do while," is twenty-four minutes long. It is made from an early piece entitled "Kardamon," which was on the *Kolor* demo tape; the original version was just over one minute in duration, and was subsequently extended to twenty minutes for the *Wohnton* installation. The audio slides between a couple of tones, while a chime-like sound stutters over the top of these. The actual CD stutter has been left behind in this track and cannot be heard, but it remains a memory as the themes in the piece jump and mistime. The track loops around the central themes many times over its course, with almost no change from loop to loop. It does not follow an obvious song structure, nor does it have any clear moments of tension or release; instead from start to finish it is in stasis. There is nothing harsh in the mix; its length suggests a type of ambient environment without peaks and troughs. Oval had at this point crossed over from pop to art: for Popp himself, Oval was now more about file management and contemporary electronic aesthetics than it was about music.[72]

Oval's sound since *94diskont* has moved away from the overt use of the skip and CD glitch; the rhythmic "dechingdechingdeching," now embedded deep in the piece, is almost covered with the noises that stream out of Popp's computer. The original skipping CDs, the actual CDs in the CD player, are a distant memory, albeit an important one. The CD skip subsequent to *94diskont* has been replaced with a sound filled with digital noise. At this point Oval has mixed back into a genre that the group helped spawn.[73]

Oval's Break and Crack

> Oval have done something subtly violent to Electronica. Once you have heard this, things sound different.
>
> —Paul Schütze[74]

In the mid-1990s there was a general feeling that electronic dance music had become tied to its tools. That is, the software had taken a firm hold on the sound of many dance-based musics, leading to output that all sounded the same, to the point where the application used to create the music could easily be discerned. This was seen as a deadlock, a dead-end for electronic dance music. As Simon Reynolds wrote in an article on the label Mille Plateaux:

> Oval resist this deadlock, or expose it, by having an "audible user-interface." In nuts and bolts terms, this involves fucking with the hardware and software that organises and enables today's post-rave electronica.[75]

Applications such as Cubase make it possible to perfectly time and lock down beats or sounds. The grid, with its vertices connected at precisely regular points, allows sounds to be snapped together at an exact intersection. Rhythm, samples, loops, and sequences all become regular in a copy–paste workstation environment. These applications might be argued to make such things too simple, creating a music that is too mechanically perfect. The ease with which this perfection can be attained through the use of these software environments then leads to the multiplication of music makers who reach a deadlock in ideas and output, simply because the technology directs them all down the same path.

It is clear, however, that many artists take the tools of music creation and push them in directions they were not designed to take. The crack exists at the point where the tools of music creation or music reproduction start to break down. Any audio application can be coopted into creating sounds well beyond what it was developed for. Cubase, or any software, certainly does not have to be used in the manner it was originally intended. This line is part of a wider argument debated during the 1990s about the relationship between new music technology and music creation. Nabeel Zuberi comments:

> In the digital age, production and reception of music are often mediated by the same reproductive technologies. Models of communication that still insist on a clear division between production and reception must be modified in the light of these shifts in mediation.[76]

Zuberi discusses an argument posited by Paul Théberge that points to the limits the production/reception situation presents for composition. The music producer has become a consumer, who is then led by consumerism and the market. This consumerism, though not new—consider, for example, guitarists and their nearly fetishistic desire for amps, effects boxes, and various models of guitars—is developed to an even greater degree. New forms of technology are products in the marketplace and are driven by market values.[77] The dependence on new and developing technologies, which exists in direct relationship to demand, has also led to a "cottage industry" that supplies new sounds and patches for digital instruments. Musicians, then, rely on these developers for new and interesting sounds, while the cottage industry itself relies on the decisions made by the large manufacturers who make the product successful and useful in the market.[78] These industries define the product's possible outputs and the range of sounds made available by the product. We have a loop whereby technology is developed for the perceived needs of the producer, whose work is shaped by the available technology is available and what it allows him or her to do with it.

In relation to the use of sampler/sequencers, MIDI, and Cubase, Popp argues:

> There is so much determinism within these programs, working with them involves so much compliance to principles that are highly critical. In a social context these technologies are mostly used in a controlling way:

> monitoring the workplace, workplace efficiency, optimising the user-interface. On-line newsgroups are full of people who e-mail back to the manufacturers saying, "We'll need this, change that," and all of this keeps them in front of their computers even longer. Our way of dealing with this is to overcome the manufacturer's distinction between "features" and "bugs."[79]

The idea that Oval was somehow making a radical break with the audio software current at the time is a simplification made by Reynolds, who does not make clear how they actually broke the applications they were using or even how they broke with the dance scene itself. Oval took Cubase and Sound Designer and used them to sequence their stuttering CD samples, looping and changing their pitch. Just how this was a radical break in the use of the software is not obvious. What was radical was the source material for their audio and how they arranged the glitching sounds to make them tuneful.

As the band's background in pop music shows, Oval's music was not a reaction to dance music or culture, nor did they develop their sound out of a dissatisfaction with dance music technologies. It is, however, also clear that Oval and especially Popp were in part reacting to the "politics of digital audio" and the processes of creating music out of digital databases.

Kim Cascone refers to the fact that what was most creative and radical about Oval was its aestheticization of the CD skip.[80] It is clear that Oval used the CD skip and glitch to great effect. Oval turned an annoying and distressing sound into something listenable, beautiful, and catchy. The aestheticization of the unwanted audio of a skipping CD player has become the basis for a generation of digital audio producers, who see Oval as the originators of digital experimental musics. While Oval continues as a solo project for Popp, it was the sounds produced in 1994 and released as *Systemisch* and *94diskont* that were the radical move. On the electronic music scene, things sounded different after these releases.

5 Disc and "Real" CD Damage

Oval is the forebear of the use of cracked media within the digital audio scene. The glitch the group caused is heard as cracking open the ears of digital audio users, revealing the usefulness of technological breakdown as the basis of sound generation. As glitch producers, Oval locked down their skipping CDs, sampling the sounds before looping and sequencing them into pop tunes;

controlling the accident is at the center of the band's practice. Future work in the use of breakdown and the crack either went to great lengths to control (sample and sequence) the breakdown, or to let it be much freer in allowing for the accident. In opposition to the planned accident, Disc represents the "gonzo" end of digital audio, taking CD destruction to a new level.

The genre of rock has at times toyed with the idea of serious damage, both to the tools of production and to the body of the performer. The guitar in particular has been bashed, smashed, and burned in the hands of many including Jimi Hendrix and Pete Townshend of The Who. The best-known example of deliberate harm being caused to the body during a rock performance is Iggy Pop who, in one particular set, deliberately caused himself to be beaten up. Other examples include numerous punk and industrial live performances, where such injury was almost commonplace.

In the scene based around the use of cracked media it was the group Disc that caused the most damage to the software of the CD system. The grouping consists of audio producers related to the label Vinyl Communications: J Lesser (Jason Doerck), Kid 606 (Miguel DePedro), and Matmos (Drew Daniel and M. C. Schmidt). The group produced three albums in 1998 before they went on to work further on their individual projects. Their interventions with the format cannot be wiped away like the marking pens used by Oval, nor simply pulled off like the tape used by Yasunao Tone. After Disc manipulates their discs they are well beyond repair. Like Christian Marclay and his use of the overabundant vinyl medium, Disc has taken to using cheap thrift-shop CDs, and destroying these valueless objects. Like vinyl in the 1980s, the CD is no longer treated as a valuable commodity: it is produced in millions of units, devalued to the point of becoming the drinks coaster mentioned on the initial release of the format.

In performance and in the studio Disc took to completely destroying CDs. The group recorded the resulting audio without the use of sampling, instead simply letting the discs play out without additional intervention. Disc used razor blades, needles, tape, and gum, sourcing their CDs from various free outlets such as promotional bins at radio stations. The group released recordings on clear vinyl on Deluxe Records, and a small number of CDs in the late 1990s and early 2000s. The releases *2 x CD* (Vinyl Communications, 1998) and *Brave New Girl* (Vinyl Communications, 1998) are notable for including over forty tracks.

Disc sequences CD glitches into tracks and "locked grooves," forming rhythmic and almost danceable tracks while still retaining the noisy and unpredictable nature of the skipping CD sounds. The group genre-bend its music as different styles are forced into humorous collisions.

Disc member Lesser has a long history with the skipping CD:

> I've been obsessed with skipping CD's since I first saw one . . . about '84. I would always try to get them from people and fool around with them. Cut them up with razor blades, putting tape on them and stuff. The first Lesser tape has an extended piece with a skipping Smiths CD. Morrissey sounding even more pathetic going "You-you-y-y-y-youyouyouyouyou-You . . . You."[81]

Lesser did not know the music of Oval at the time he initially played with skipping CDs. He had simply been making tape recordings of various skips for his own listening pleasure—a twist from the displeasure most find in the sound.

The clear difference between what Oval was doing and Lesser's approach is one of simply letting the CD skip and stutter, of listening to the accident. Lesser is happy to hear the CD glitch documented, without additional input from himself and without making the tracks into songs or pieces.[82] Lesser lets the sound of the stressed CD system be itself—more akin to John Cage's approach than the digital audio scene. Lesser is here closer to Yasunao Tone than he is to Oval.

Whereas Tone was interested in the use of the skipping and stuttering CD as a method of indeterminate composition, Lesser hears textures and rhythms. The San Francisco scene he is part of is based around broken and cut-up beats. Kid 606 and Matmos, for example, take beats from different genres and mash them together in their music. We hear from track to track, and sometimes in the same piece, breakbeats, hip-hop and house rhythms, brass bands, and even folk. Lesser heard the stuttering CD as a free rhythm and a possibility for making new uncontrolled rhythmic patterns. He points out that he

> wasn't interested in making lovely sounds or full songs out of CDs. . . . I just wanted to make recordings of the CDs for myself. . . . I liked how free-jazzy the rhythms were . . . it's not about the tonality for me . . . [it's] strictly rhythmic.[83]

It is possible that this “jazzyness” comes from the high-end tick reminding Lesser of the high hat used in some jazz, as well as the polyrhythms that occur during the CD stutter. The skip of the CD is certainly open for rhythmic use, and Oval did this but in a more structured way.The erratic nature of the actual skip and stutter of a glitching CD might be too risky for tracks based in pop structure, but for Lesser they were a ripe method for a type of found sound, found rhythm.

6 Conclusion

Yasunao Tone, Nicolas Collins, Oval, and Disc represent four different approaches to the use of the skipping and glitching CD in experimental sound composition and creation. The process for Tone is bound up in his ongoing interest in chance and indeterminate techniques in composition. This approach has its history in the avant-garde and Tone’s own history in the experimental arts in Tokyo in the 1960s. Collins works with the production of “idiosyncratic” electronics, tampering with preexisting equipment to produce his compositions and performances. His work with electronics and CDs is not so much about the destruction of the technology as it is about the manipulation and even redesigning of existing equipment. His interest is similar to Tone's at the point where catastrophe meets chance. Both composers are interested in chance and the new and original outcomes created by the process.

Oval's audio represents the planned use of glitches. A chance occurrence occurred in the past and now has been used and reused for samples in their heavily sequenced work. Oval's music is never actually played live in any real sense. An Oval work is not an indeterminate score or composition; rather, the works are finished and sealed pieces of pop. This process owes more to the pop music process of recording than it does to twentieth-century composition. Disc presents a much more haphazard approach to the damaged CD glitch. The CDs are broken completely in performance and the band has little regard for the medium itself. They thrive on a destructive and an almost punk-like approach to both electronic music and the use of technology.

In the late 1990s, in the genre that Oval "double-clicked," producers, especially those involved with the German record label Mille Plateaux, increasingly focused on the accident and chance in a digital environment. Digital technologies were pushed to their limits and the distressed sounds were recorded, to be later cut and pasted into a final sequenced track. In addition to the sample and sequence technique, many laptop musicians, including those associated with the Austrian record label Mego, have taken to pushing things further, opening their applications and code to invasion from disparate data to produce an unknown and unforeseen audio. There is no doubt that Tone's, Collins's, and Oval's use of the relatively new CD technology has been, and continues to be, an important influence to producers working in the area of expanded techniques and experimental sound and music.

4 tactics, shadows, and new media

1 The Everyday: Michel de Certeau's Tactic and Cracked Media

A clear and important commonality in the practices of cracked media is the everyday nature of the technologies utilized. These technologies are ubiquitous—almost everyone has some sort of playback technology in his or her home, whether it be a sophisticated hi-fi or a small mono radio. It could be argued that, quite simply, these sound technologies were used creatively because they were easily accessible and close to hand. The factors of accessibility and affordability do account for some of the practice, but they do not explain why, for example, Fluxus-related artists broke and destroyed current and often valuable technologies in their work. This equipment is everyday in nature; hence a possible framework for understanding the practice of cracked media may be found in the theories of creative practices of the everyday.

The discourse surrounding the everyday and the practice of the everyday in the arts has centered on a number of key French texts written in the second half of the twentieth century, including Guy Debord's *Society of the Spectacle*, Jean Baudrillard's *Simulations*,

Michel de Certeau's *The Practice of Everyday Life*, Pierre Bourdieu's *The Field of Cultural Production*, and more recently Nicolas Bourriaud's *Relational Aesthetics* and *Postproduction: Culture as Screenplay*.[1] These texts provide models for creativity that focus on how the everyday can itself been seen as a site of politically charged activity.

Cracked media are simply redirected electronics. The producers who exploit these redirected electronics take the tools of playback and redirect them to transform and create sounds for unique and/or indeterminate outcomes. These often commonplace or everyday technologies become performance tools that extend the sound field and modes of sound performance. Michel de Certeau's concept of the "tactic," as outlined in his *The Practice of Everyday Life,* demonstrates the implications of the tactical application of everyday technologies. De Certeau's theories are often taken as political in outcome; for example, the concept of tactics was used by Hakim Bey in his conceptualization of the "Temporary Autonomous Zone" (T.A.Z.).[2] This is a zone in which for a brief period the laws and rules that govern the state are overcome and the zone for this limited period is governed by its own new rules. The key was that in employing the tactic it was understood the although the zone would not last for an unlimited period, it could be fully enjoyed in the here and now. Political groups such as Reclaim the Streets utilize the Temporary Autonomous Zone in their anti-car protests-turned-street-dance-parties.[3] A recent poetical example is the work of Graffiti Research Lab and their *Throwies*—small LED (light-emitting diode) devices that are thrown on metal walls and stick there because of a small magnet.[4] A group of people hit a specific site, such as a wall, public sculpture, or a tram, with hundreds of these colored lights. For a brief time the austere structure is made colorful and bright.

This overtly political interpretation of the tactic is, however, not the only possibility. The tactic can also take the form of a nonpolitical action; this is important when the theory is applied to the contemporary sound and experimental music scene, which is not known for its overt political intentions. The tactic is a way of living and offers an approach to existence that relies on making life easier, or better, by making do.

De Certeau's Tactics and Modes of Operation

In *The Practice of Everyday Life* Michel de Certeau investigates the ways in which cultural users operate, providing pathways for further research. He asserts:

> This goal will be achieved if everyday practices, "ways of operating" or doing things, no longer appear as merely the obscure background of social activity, and if a body of theoretical questions, methods, categories, and perspectives, by penetrating this obscurity, make it possible to articulate them.[5]

De Certeau is suggesting a return not to the individual, but instead to "modes of operation or schemata of action, and not directly the subjects (or persons) who are their authors or vehicles."[6] The actors in this are concealed in "culture" within the terrain of the "consumer." These consumers then poach the property of others as part of their daily life, using it, in this case, for creative purposes.

In the chapter entitled "'Making Do': Uses and Tactics," de Certeau outlines the difference between the "strategy" and the "tactic." He explains:

> I call a *strategy* the calculation (or manipulation) of power relationships that becomes possible as soon as a subject with will and power (a business, an army, a city, a scientific institution) can be isolated. It postulates a *place* that can be delimited as its *own* and serve as a base from

> which relations with an *exteriority* composed of targets or threats (customers or competitors, enemies, the country surrounding the city, objectives and objects of research, etc.) can be managed.[7]

The strategy is that which is carried out with a center or a place, with autonomy, whether real or imagined. With place, perhaps literally land, comes power and also legitimization of rules and laws.

The tactic, on the other hand, is carried out without place and therefore without any delimitation of exteriority. The tactic is one of opportunity, carrying out actions "blow by blow":[8]

> It takes advantage of "opportunities" and depends on them, being without base where it could stockpile its winnings, build up its own position, and plan raids. What it wins it cannot keep . . . It must vigilantly make use of the cracks that particular conjunctions open in the surveillance of proprietary powers. It poaches them. It creates surprises in them.[9]

Tactics are not methods to cause a revolution; they are not set up to overthrow the power and its strategies. Instead, they are there to make life a little better and perhaps a little freer. Of key importance to understanding the difference is de Certeau's idea of "making do." This is achieved by taking what is owned by those with place and territory and using it in one's own way. This will not change the intended purpose or the strategy set for the object in any real sense; any winnings cannot be stockpiled, as there is simply nowhere to keep them; but for a moment there is certainly freedom to use them as one desires.

This model of power aids in understanding the workings of the "everyday" in contemporary creative practice, as musicians and artists quite literally make do with objects, images, and ideas that are found, tampered with, borrowed, or stolen from the everyday. The producers who use cracked media take everyday home entertainment equipment and music storage devices and use them to create something new. The original intention of transparently reproducing a prerecorded piece of music is thus snatched and redirected toward an original and creative act.

Christian Marclay discovered vinyl records literally lying in the gutters of New York; he scavenged this rubbish, reusing it to produce original sounds and artworks. Here he is using surplus goods from the everyday to produce new and original works. Milan Knížák became bored with the few records he could afford to own and so came up with a way of reusing them to create new music by slicing them into segments and reattaching them to form new compositions. Here he is applying tactics to the technology. Yasunao Tone takes a CD of Debussy's music and attaches tape to its surface, manipulating an everyday recording, produced in thousands of identical units, to create an indeterminate performance. These practices can be seen as the tactical use of everyday technologies. De Certeau argues:

> Many everyday practices (talking, reading, moving about, shopping, cooking, etc.) are tactical in character. And so are, more generally, many "ways of operating": victories of the "weak" over the "strong" . . . clever tricks, knowing how to get away with things, "hunter's cunning," maneuvers, polymorphic simulations, joyful discoveries, poetic as well as warlike.[10]

Unlike the strategic modes of production, the use of cracked media is outside of the "correct" means of creation. These producers use what they have and what they can find, often for little or no cost, to create new sounds and new works. Oval used the local public library, borrowing already used and damaged discs to sample for their glitch pop. John Cage used simple objects such as toothpicks and pipe cleaners to act as needles in his modified cartridge heads; and Michael Graeve pulls rasping noises from numerous found, borrowed, and discarded turntables. Making use of "clever tricks," practitioners of cracked media exploit their tools outside of their everyday strategic uses. This practice demonstrates what de Certeau calls the "joyful discovery," which can be witnessed in, for example, Nicolas Collins's discovery that when the mute chip is removed from a CD player all the sounds of the functioning of the device are played back as audio.

As previously argued through the critique of Attali's "repetition" phase, the use of cracked media contradicts the purely negative effects of recording on music practice. In the terms of the current discussion, the recording industry itself is the strategy, generating homogeneity through the mass reproduction of music. A possible reply to the criticism of Attali's argument comes in the form of his own prediction for a then-future phase of music that he calls "Composing." He sees in music a prophetic shift away from consuming, away from owning recordings of music, to the creation of music by the individual for the individual. This predicts the "very death of exchange and usage in music, the destruction of all simulacra in accumulation" and the "bringing about [of] a renaissance."[11]

For Attali the way forward is not simply a new music but a new way of making music:

> Doing solely for doing, without trying artificially to recreate the old codes in order to reinsert communication into them. Inventing new codes, inventing the message at the same time as language. Playing for one's own pleasure, which alone can create the conditions of new communication.[12]

Attali's "composing" phase is very close to the practices of cracked media and can be equated to and extended with de Certeau's tactics. By situating the crack and break as a tactic we can also place these practices within a wider cultural framework and also within a wider arts discourse. In the 1990s de Certeau's notion of the tactic was used extensively as a way to explain visual art practices that made use of the everyday and seemingly mundane subjects. Numerous examples of this practice exist in contemporary art, including: Ceal Floyer's *Garbage Bag* (1996), which was simply a garbage bag placed in a gallery to form a minimal intervention;[13] Clay

Ketter's *Surface Habitat for Appliance* (1997), which are IKEA-like sculptural kitchen shelving units;[14] and the relational works of Rirkrit Tiravanija, who produces situations in which social activities occur, often within the gallery environment.[15] In his catalog essay for the 11th Sydney Biennale entitled *Everyday*, Nikos Papastergiadis writes:

> From the Dadaists and the Surrealists, to the Situationists and Fluxus movements, investigations into the meanings of the everyday in modern art have been ongoing. At their centres has been a critique of the dominant forms of urban life in particular.[16]

As in everyday practices within contemporary art, the use of everyday playback equipment forms a critique of the dominant modes of use of the technologies. Modes of use for playback equipment, based on its imagined transparency, and the convention of the closed media object are critiqued by the practitioners of cracked media. This desire for the perfect copy is turned on its head, as the crack and break draws attention to the materiality of the media through a critique of its form.

2 The Shadows of Technology: Paul DeMarinis and Kim Cascone

The use of cracked media provides original, transformative, generative, and indeterminate sounds. These sounds are filled with noise as they pop, tick, skip, and glitch throughout the work, and they form the content of the compositions and often the structure. Yasunao Tone's "Wounded CDs," for example, form both the sound content and the indeterminate structure of his performances. Work such as that produced by Otomo Yoshihide uses the turntables themselves to produce the source material and structural elements for his improvised performances, while the manipulated electronics of Nicolas Collins can also be understood to provide sound material and compositional structures. Oval samples indeterminate CD skips and stutters, then very deliberately produces tracks through methodical sequencing; John Cage in *Cartridge Music* uses cracked turntable hardware to provide the support and framework of the performance of the composition. All of these approaches create novel sound and structural possibilities. Yet beyond the urge for an expanded sound palette and new structures for composition, what might some of the rationale for these activities be? Why have these techniques been employed, beyond their generative novelty, and what is the significance of the crack and break?

Put another way, what is the sound of the crack and why has the often noisy misuse of technology led to so many productive outcomes? What might practitioners who use, for example, malfunctioning and modified turntables be hearing? What are they aiming to hear as they step back from the traditional use of their equipment? One possibility is that they are looking toward the gray regions of sound, hunting in the shadows for the ghosts of signals left behind, the shadows themselves becoming the focus rather than the body that throws them. That is, what we have is the deliberate framing of audio that was not meant to be heard, audio that is usually a mere glimmer hidden behind the brightness or reality of the "real" or "desired" audio output. These producers extract sounds that had previously gone unheard, or had simply been relegated to the far background or the rubbish bin of audio culture.

Paul DeMarinis and Kim Cascone offer two similar approaches that can be used to illustrate the reasons for the use of cracked media. Both are practitioners, but here I am more interested in the theories they propose than the work they produce. DeMarinis will be examined for his reading of phonographic effects and the idea that the invention of the phonograph not only foregrounded the possibility of recorded sound, but also opened the gates for a flood of extra and unplanned effects. Cascone also understands the practices of "postdigital" music as making use of additional, and normally unwanted, outcomes of the digital studio, and it is his idea of detritus that most closely aligns with cracked media.

Paul DeMarinis is a multimedia artist who has been working with installation and performance since 1971. In relation to his work *The Edison Effect* (1989), an installation with a laser trained on a vinyl LP, DeMarinis discusses three sets of sounds that were heard when early recordings were played through the phonograph. The first was the intended sound of the recording, the performance. The second was environmental sounds, the sounds of background noise such as birds, people talking nearby, and so on. DeMarinis points out that the phonograph helped make people aware of these sounds and spawned the field of environmental sound art. The third set of sounds discovered by the phonographer was the "scratching and rumbling of the machine itself, inextricably registered along with the sound it was supposed to faithfully preserve. Unlike the environmental sounds, these sounds would not exist if the recording had not been made. I call this the shadow of the technology."[17] The sound of the rumbling motor that drove the device, for example, registered on the final recording. Wax, shellac, or vinyl itself has a distinct sound; even CD playback and its pinpoint ones and zeros bring with it its own sound, distinct from the wave patterns of analog playback. DeMarinis further describes the sound on the surface of the wax:

> the texture and grain of the wax has its own raspy voice, a voice that sang along with every diva and accompanied every chance sound passing by the microphone. Surface noise, channel noise, the song of long ago and far away, presented a gift in disguise to the recordist and artist alike. This noise is an audible indication that information is being sent. In effect this "noise-floor" is the sound of silence of any given channel.[18]

DeMarinis adds a fourth dimension of sound heard when a record is played back that he calls "autobiographical." These sounds are formed after the record has been bought and are caused by the wear and tear of playing the record—the scratches and dust that cannot help but become ingrained on the disc's surface.[19] Record collectors soon find that the very act of playing their precious objects causes their slow destruction. With each subsequent playing the grooves are a little bit more damaged by the stylus, no matter how good the quality of the diamond or how careful the user is by keeping the disc clean and placing the tone arm down as gently as possible. In time the recording itself becomes muffled, not to mention marred by the growing number of pops caused by dust, lint, and scratches. Also, those who used CDs shortly after their initial release soon discovered that the rhetoric of permanence did not hold true in practice: CDs that were not handled carefully soon developed glitches, skips, and stutters, making the playback of the CD frustrating or even impossible. As has been shown, this autobiographical sound was significant for practitioners of the crack and break: for these artists it was through the daily use of recordings and their playback technologies that they developed an interest in and aesthetic based on its breakdown.

In these recording media, the "sound of silence" is thus actually extremely noisy, full of hum, hiss, and the cracks and sparks of static. The grooves of records have been filled with music and sounds that aim to drown out any surface noise, pushing it well under the volume of the "real" object of the recording intended for listening. This noise was then further suppressed with the development of recording studios and digital recording techniques, both being designed to remove all environmental sound and the noise of the recording process itself. However, DeMarinis's third and fourth dimensions of recorded sound are never fully removed from a recording; even in the pristine digital studio these sounds, at infinitesimal levels, are immediately below the surface and perceivable, even if barely so.

The shadow effect can be heard in its literal use in work by turntablists practicing in the 1980s and '90s. For example, Canadian turntablist Martin Tétreault generates sounds in performance by placing the tone arm close to the turntable's motor. The hum of the motor is joined by other "pure" sounds of the turntable such as the sound made by putting the tone arm directly onto the platter. Using no records at all, the sound of the machine itself is brought directly to the fore to become the content and material of the work.[20] In the hands of Tétreault, the turntable becomes an abstract noise-making machine: the memory of recorded media is completely discarded and replaced by noises picked up by the cartridge head, which has been modified in its use to become a contact microphone. These sounds are directly linked to DeMarinis's third dimension; they are sounds that would not have existed without the invention of the phonograph. Here Tétreault is quite directly performing the technology itself.

In a similar way, Michael Graeve performs the phonograph in a manner that draws out this third dimension or shadow effect. In performance, among other turntable-specific effects, he causes the turntable to feed back by placing the stereo system's speaker very close to the stylus. Otomo and Akiyama Tetsuji also draw on the shadow effect and go beyond it, as they both use the phonograph without records and seek out the effects caused by extending the everyday functions of the turntable to the point where it no longer resembles a phonograph at all; it simply becomes a rotary sound-making device.

Other practitioners of the crack, such as Marclay, foreground DeMarinis's fourth, autobiographical dimension, focusing on the scratches and marks on vinyl records. Marclay does this very deliberately through allowing the objects to be marked by not covering them when they are not in use, or by placing them on the gallery floor and allowing patrons to walk over them. Oval too uses a similar approach, seeking out CDs in their local library that had been already damaged through normal use, and recording the skipping and glitching effects these marks caused when the CDs are played.

Like DeMarinis, Kim Cascone works with background noise and the by-products of the processes of recording. Cascone is a sound maker and writer working with the new tools of computer-mediated sound production. In his text "Residualism," Cascone discusses the process of "removing a signal until all that's left is its ghost-signal or the artifacts thrown off by the signal."[21] Within the environment of the digital studio, many producers have left the clinical high-tech world of digital production to hunt through the detritus of the sound-making process, hearing the off-cuts as material for aural exploration. Cascone explains:

> most of the digital detritus gets noise reduced, multi-band compressed, brickwall filtered out . . . but we now find ourselves immersed in the next wave of digital audio. . . . new tools have emerged that are capable of exploring both the signal and what's behind it. . . . we are able to descend into the "abyssal zone" of the noisefloor and use what we find there as material for exploration.[22]

The material explored by artists working in postdigital audio, according to Cascone, is the subtraction of the signal: the layer that is left after the "proper" meaning of the communication has been removed or peeled back, the original signal attenuated, and the background amplified. Noise-reduction software has removed noise from the file, metaphorically moving it to the digital trash where these producers have scavenged it, finding use in the waste products of the digital studio. Producers working in the area make use of sound-production and sound-creation software to search through the "noise-floor," looking for useful off-cuts and interesting sounds. These software applications act as metaphorical microphones for the rehearing of undesired audio:

> we have only recently begun to consider the gauzy veil of hums, clicks, whirs and crackling as worthy of our attention. . . . data-mining the noise-floor is today's alchemical pursuit of turning bits into atmospheres.[23]

This noise-floor is the hum at the very bottom of the mix, the tape hiss that has almost but not quite been covered with the content of the recording. The noise-floor is related to DeMarinis's third dimension: it is an effect of the technology that would not have occurred without its invention. These sounds are usually heard as an annoyance, at the very least, and at worst they disturb the signal. An example of this is the heavily played vinyl record. The signal interruptions, the pops and ticks, reach a point where the record is no longer of any use, in the traditional sense; with the musical content covered by these noisy distractions the vector of communication is broken. Imagine, however, shifting our focus from trying vigilantly to hear the real content of the record, the intended material of the sound recording, to listening to the ticks and pops as the content. This shift changes background to foreground and vice versa.

This shift in focus occurred in the underground scene of digital audio in the second half of the 1990s, spurred on by the home digital studio, but the idea itself has been a key element of cracked media from the beginning. The story, for example, of the use of broken and cracked turntables often starts with the producer listening to his or her much loved record collection that has been worn down and damaged through multiple playings. The sound producer begins to attend to and actually hear the ticks and pops on his or her records, and finally the scratch that causes the needle to jump its groove, as sounds with their own aural interest and their own possibilities as material for audio practice. This was the case with Knížák as well as Marclay, both of whom heard these effects as aurally interesting and went on to force these playback events to occur through the modification of the record.

In "The Aesthetics of Failure" Cascone describes an aesthetic that is focused on the notion of "detritus," "by-product," and "background" in digital audio. He states:

> The basic composition of "background" is comprised of data we filter out in order to focus on our immediate surroundings. The data hidden in our perceptual blind spot contains worlds waiting to be explored, if we choose to shift our focus there. Today's digital technology enables artists to explore new territories for content by capturing and examining the area beyond the boundary of "normal" functions and uses of software.[24]

The producer working in the digital studio could not help but come across digital failure. The postdigital producer inevitably comes into contact with unwanted or unplanned "glitches, bugs, application errors, system crashes, clipping, aliasing, distortion, quantization noise, and even the digital hash contained in the noisefloor of computer soundcards."[25] Similar to those who spend time with recorded media, the practitioner of the digital environment is soon forced to deal with the failure of the medium. As we have seen, these side effects of digital sound production were then used by some as material for a developing postdigital aesthetic.

Another possible way to hear digital detritus is as the off-cuts of the production process. The scraps of sounds lying in wait to be removed and trashed are discovered by the attuned aural discoverer who hears them as ripe for exploitation. Cascone tells us that filmmakers in need of ideas have often rummaged around the cutting-room floor looking for new material.[26] With this in mind, in 2000, Cascone initiated a project entitled *abstractMachines*, based on the sound files that formed a previous work *pulsar studies*, a series that was distributed on the Internet site http://fals.ch as one-minute MP3s.[27] Cascone sent these to a number of artists including Aube, Autopoieses, CD-Slopper, Karkowski, Kid 606, Neina, and Pimmon. They were asked to remix these sound bites with the intention that new works be created by others using the sound bites and off-cuts from Cascone's output, the sound becoming the inspiration and source for the new works. The detritus of the digital studio is thus picked up and reused, and old material, the waste from previous recordings, is employed to create new audio works.

The move to the shadows or the ghosts of sound production leads swiftly to the deliberate crack or break. Once the ears have been honed to the third and fourth set of sound outcomes from recording and the phonograph itself, the producer soon needs more. He or she can no longer hear enough in the microscopic regions, in the darkness of the shadows, the translucence of the ghosts and off-cuts of others' productions. Desiring more, the practitioner soon forces the sounds he or she is looking for, pushing and twisting the audio until the deliberate accident occurs. Oval, for example, began to force the skip by adding marking pen to the CD substrate, while Thomas Brinkmann literally cut into the "runout groove" of the record, causing a deliberate locked groove that would produce regular ticks to form an error-driven drum pattern.

As previously discussed, the work in this area not only pursues extended sound practices, but also draws us into the actualities of the media, its materiality, and its everyday uses. It was argued in the introduction that the technologies of sound reproduction are understood to progress toward an impossible transparency. As Jonathan Sterne writes: "Sound fidelity was, ultimately, about faith and investment in these configurations of practices, people, and technologies. It posited the technology to reproduce sound as a vanishing mediator."[28] In the practices of cracked media the material support for these technologies is foregrounded and exposed, and as such the dream of transparency is deemed undesirable. The weaknesses of these structures are exposed as the technologies are pushed and pulled into creating unintended sounds. This is nowhere more clear than in the work of Collins, who accesses the internal mechanisms for reading data and allows the data to audibly play. The various notions of perfect, clean playback, unaffected by mediation, are exposed and exploited by the practitioners of cracked media.

The practitioners of the crack and break focus on the materiality of the medium and as such reveal the infrastructure of these playback technologies, but what lies beneath the surface? It is clear that there are technological workings and that these workings can be tampered with, manipulated, cracked, and broken; but there is also a cultural layer that is broken through here. As Sterne observes, it is the social and the cultural that allows the imperfections of the technologies to go "unnoticed." The cultural practice of listening to these media objects explicitly requires that the listener ignore or block out those elements of the media that would return him or her to the actualities of the mediated experience. That is, the listener is required to buy in to the myth of transparency. The practitioners of the crack and break expose these layers, showing them to be false. As Mitchell Whitelaw argues in his article "Inframedia Audio," these types of practices open a space within the cultural mainlines. Whitelaw posits that in the current situation of expanding media technologies, it is the infrastructures of these technologies, and not media content, that are of key concern.[29] In other words, these works are fundamentally about the technologies themselves.

However, the nature of the crack and the break suggests that once the materiality of the technology has been exposed, there is little further development that can occur. The aestheticization of the CD glitch, for example, cannot be pushed further (or the CD will simply stop playing altogether), and as such the approach has a limited number of audio or structural outcomes. Soon after Oval's initial experiments with skipping CDs, issues emerged about the future for a movement based on glitch and the breakdown of technology. How many times can a skipping CD be heard as unique? Given the limited nature of the CD glitch, why does the turntable continue to be used in this way and continue to yield novel and unique outcomes? One possibility lies with the very materiality of the turntable. Given its analog nature it can be attacked, thumped, and radically manipulated and still play. The CD by contrast can stand only a limited amount of abuse. If the CD is scratched wrongly, if too much code is covered or corrupted, it will simply not play. There is a very fine line between a skipping CD and a mute one. Also, the CD player has a close and exclusive relationship with its playback format: a CD player will only play CDs. The laser, unlike the tone arm, cannot be applied to other objects and still produce sound.

The use of the phonograph in the twenty-first century provides an alternative to the immateriality of digital media. The limitations of CD technology are not a problem for the much sturdier turntable. The needle will attempt to plough through the most damaged records and will attempt to play whatever it is placed on, even the platter if a record is not present. The possibilities for extended use of the turntable are still being mined as seen and heard on a DVD release entitled *Multiple Otomo* (2007). The DVD features a range of short sketches by Otomo. We witness up close a multitude of his approaches to the use of the turntable, including: the gentle use of a stretched rubber-band and a lump of aluminum foil placed on the platter; metal springs replacing the stylus that are in turn attached to colored vinyl or bowed with a violin bow; and the pouring of pink liquid over a turntable while it attempts to play a record (the phonograph does a good job of ignoring the foreign substance and manages to play the record as if nothing unusual was occurring). Because of the phonograph's analog robustness it has many more possibilities for unintended use, and as such some 130 years after its invention new ways of expanding and abusing it are still being discovered.

3 Cracked, Broken, and New

New media has at its core the ability to transform digital code into new objects. We take images and sounds from various sources and file types (some found, some built from scratch) and blend them together into a final product such as a Flash interactive. These new media objects are not fixed, and they can be developed and altered in an infinite number of ways. Lev Manovich argues:

> Old media involved a human creator who manually assembled textual, visual, and/or audio elements into a particular composition or sequence. This sequence was stored in some material, its order determined once and for all. Numerous copies could be run off from the master, and, in perfect correspondence with the logic of an industrial society, they were all identical. New media, in contrast, is characterized by variability. . . . Instead of identical copies, a new media object typically gives rise to many different versions.[30]

It is my argument that the practitioner of cracked media approaches old media in a manner that mirrors the attributes of contemporary new media. New media often relies on the use of already created objects or artifacts, such as images or sampled sound libraries, to produce the final product. These products are themselves available for further use, and as Manovich states above, this variability is a hallmark of new media. Those who use cracked media do not passively consume media; instead they transform the products of old media to generate new works. What was considered sealed and complete has been taken apart and transformed into new variations of the original or even complete unique outcomes.

Records were originally created to be consumed in a fixed manner. The listener of a recording is imagined to be seated, perhaps in his or her living room, and was meant to listen to the record from start to finish to gain the most from the experience. These recordings had been produced as a final and sealed product, and it was expected that the only changes that would occur to this product were from general, undesired wear and tear. The development of compact disc technology held the promise of permanent copies of a recording. This is in fact quite possible if individual CDs are treated with great care; but CDs also succumb to general wear and tear, leading to skips and glitches occurring during playback. These objects where produced to be played and listened to in a certain way, and they were certainly intended as final and complete, but it is also clear with these types of media that this was never quite possible. The media did wear out and develop marks and scratches simply by their use, and these marks were different from copy to copy as they reflected the unique "life" of each one. Mass-produced recording then can become personalized, and in fact it is quite possible that these individual examples may be productive. What was intended to be a closed and final product is redirected toward original outcomes, whether through the cracking of the recording media or the manipulation of the playback technologies. This sort of freedom runs counter to the mainstream ideologies of the industrial and mechanical era and can be seen as yet another example of the creative counterculture that emerged in the 1960s.

Manovich links the logic of old media and new media to the culture and society in which it was formed, arguing:

> The principle of variability exemplifies how, historically, changes in media technologies are correlated with social change. If the logic of old media corresponded to the logic

of industrial mass society, the logic of new media fits the logic of post-industrial society, which values individuality over conformity.[31]

The "logic of industrial mass society" was shaped by the belief that everyone should experience life in an equal and democratic manner. Media was assembled and distributed in massive runs of identical copies that were then to be consumed equally and passively by the members of that society. For Manovich our approach to media has changed in the postindustrial society. In this society, "every citizen can construct her own custom lifestyle and 'select' her ideology from a large . . . number of choices."[32] In terms of the music industry, this is demonstrated through the popularity of technologies such as iPods and the use of the program iTunes. These technologies allow users to construct the soundscape to their lives in public, in the car, and in their home.

The approaches taken by the practitioners of cracked media can be understood to preempt future shifts in culture and society. If we are currently in a society that seeks the singular and individuality of media, that is, media arranged by us and consumed when we want it—YouTube instead of commercial television, blogs instead of best sellers—then the manner in which old media was taken apart to produce new individual media is a precursor to current societal trends. We increasingly take for granted the ability to actively engage in the media; but when Knížák started to permanently transform his records for his own interest, he did something new and uniqute. Such practices are now routine, although typically less destructive. We do not think twice about taking various pieces of music from discrepant sources and bringing them together in playlists, mixes, or even Podcasts. It is also common to take a piece of music and remix it into something that better suits our tastes, offering an alternative (whether official or not) to what was originally intended.

Although cracked media can be understood as a precursor to contemporary approaches to new media, it is also important to note that the crack is a continuing and productive approach in the domain of contemporary digital production. We are, that is, in what Kim Cascone describes as the postdigital environment, in which deliberate errors are used to produce media that turns its back on the clean, methodical, and calculated digital studio. This approach is a common part of contemporary practices and can clearly be witnessed through its extreme and early manifestation in the work that surrounded the Austrian music and media label Mego. Writing from the time resonates with much of the earlier cracked media discussed in this book. For example, in a press release from "The Cyberslag Foundation," Jeroen de Boer writes about the Viennese trio Radian: "'Glitch,' a denomination for music derived from the imperfections and defects of electronica, has been tried and tooted to the extreme by a variety of Mego-artists. Scratches, crackles, high-pitched tones and a lot of space typify [the] genre."[33] Tim Owen, writing about the duo and Mego label founders Rehberg & Bauer, describes their music as being similar to that of Oval's in that it utilizes a "gorgeous array of digital glitching, the raw thrum of electricity, and the crackle of static."[34] Mego-related producers, including Farmers Manual, Fennesz, Hecker, and Rehberg & Bauer, created music on PCs and laptops that was focused on the inherent errors and break downs of digital production. In doing so they demonstrate the similarities between new media practices and old media.

In addition to noisy audio, Mego also developed noisy imagery in their CD packaging, web presence, and media projections during live performances. The imagery, often designed by Tina Frank, contained granulated and distorted graphics and text. The digital, coded look of the work has close links to cracked media and is a good example of the strategies in practice in recent digital production.[35] This book has been focused on cracked media in sound production, yet there are a multitude similar and closely related practices in the visual domain.

At the turn of the twenty-first century numerous web- and e-mail-based projects were initiated that filled the clean digital environment with noise. A prominent example of this type of work is produced by Jodi (Joan Heemskerk and Dirk Paesmans).[36] In a review for *Artforum* Margaret Sundall describes these artists as the "Dadaists of Internet art": "Like those early members of the avant-garde, their work employs strategies of rupture and subversion to create an estrangement effect, jarring viewers for a moment out of their everyday lives online."[37] The website displays numerous, often crudely animated web pages filled with seemingly broken text. Their work images the imagined internal workings of the Internet and breaks with the expected logic of the Internet's interface as pages switch randomly rather than logically with the click of the mouse. These pages at times mimic crashing operating systems or cause the user trouble by continually opening new blank browser pages until the user is forced to quit the browser.

In the work described above discrepant data forms converge in the computer-mediated environment. Contemporary new media interest in convergence and the ability for data to flow across media is directly critiqued by cracked medial practices. The contemporary use of analog equipment, specifically the turntable, suggests a new approach toward media materiality and new directions for critiques of dominant media forms and their functions. These practices remain in close dialogue with new media and digital practices, and offer critiques of the assumptions of new media history and theory. Cracked media brings us back to the material of old media playback technologies and the dialogue between these material objects and the nonmaterial nature of data.

Australian artist Andrew Gadow produces audiovisual work situated at the point of breakdown of sound and image. In a manner similar to Otomo or Graeve, Gadow critiques "new media" using "old media." His work *Techne—50 hertz* (2005) is typical of his practice as it utilizes the discrepancies of analog synthesis to generate sound and image. Gadow generates video images with a Fairlight visual synthesizer and in turn feeds the image signal into a Moog synthesizer, which interprets the signal as sound.[38] This sound signal is then fed back into the Fairlight, which interprets the sound data as image, creating a feedback loop. The outcome is a noisy image combined with the dominant sound of 50 hertz electricity hum.

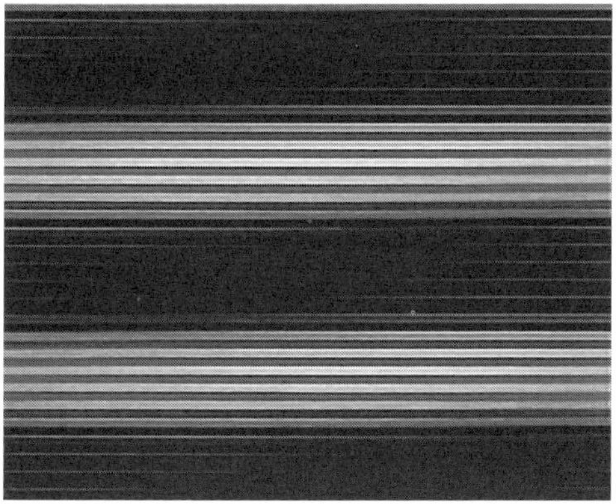

4.1
Andrew Gadow,
Mooglighting. Video still,
2003–2004.

The work of Gadow is an example of the contemporary use of old media, analog synthesis, to create a work that uses analog technologies to critique assumptions made toward digital technologies. Discrepant data is used to create images out of sound data and sound out of image data. The work is then read within the now dominant digital context, twisting the logic of the postdigital aesthetic with a return to old(er) media. Similar approaches have occurred in experimental audio since the early 2000s, including the work of Otomo and various other musicians from the Tokyo scene who have returned to older technologies and redirected their functioning in on themselves. As in Otomo's feedback-focused performances, these technologies are forced to play their internal mechanisms, the effect leading to feedback loops that are at the edge of controllability.

The reuse of the turntable can be understood as a reaction to the digital. Circumventing the discussion of the malleability of data, these artists are no longer part of the discourse of new media and its convergence. By not using new technologies, they are, however, indirectly critiquing dominant modes of practice and use. The employment of older technologies in contemporary practice draws our focus in on the materiality of the media and its intended and unintended uses. While these producers escape from directly addressing the various issues of digital production, they offer a powerful alternative view in the process. The critique brings to the fore issues surrounding materiality, transparency, and newness itself as an imperative of "new media."

Have we, in terms of the work of contemporary turntable experimenters, come full circle: has cracked media, having prefigured current approaches to digital, now turned back to its materiality? Musicians such as Otomo and Akiyama have directed their turntables to produce sound well beyond any use for the record disc, even to the point where amplification and electrical current are switched off. This approach can be read as divergent from current digital new media work and as such as forming a new thread of practice outside of new media. The scene might, however, have reached its endpoint, as the materiality of the turntable has, it would seem, given up all it has. What we have left is the remnants of the turntable, with external sound-producing objects scattered over its surface. The turntable itself has disintegrated.

notes

introduction

1. The event was performed at impermanent.audio, April 7, 2002. Images of the performance can be found at http://impermanent.info/impaud/archive/020407/images/ (accessed June 2008).
2. The year 1994 is given as the starting date, as this is the year the group Oval released its influential CD *Systemich*; the year 2000, as well as being the turn of the century, marks a downturn in experimentation in the area and a shift to the legitimization of the practice.
3. Rob Young, "Worship the Glitch," *The Wire* 190/191 (1999): 52.
4. Dean Roberts, liner notes to *All Cracked Medias* (Mille Plateaux, 1998).
5. MP3 or MPEG-3 audio is a data-compressed digital audio file format. In terms of this discussion, the format is important as it allowed music makers to send their work directly to other interested parties through the Internet and e-mail. CD-R, which is a CD format that allows for audio to be directly burned to disc from a computer, had a similar effect. Audio could easily and cheaply be burned and given or swapped with other interested parties.
6. Jacques Attali, *Noise: The Political Economy of Music* (Minneapolis: University of Minnesota Press, 2002), 133–148.
7. There is a solid body of writing on various issues created by performing live with a laptop. See in particular my paper in a special laptop issue of *Contemporary*

Music Review: Caleb Stuart, "The Object of Performance: Performativity in Contemporary Laptop Performance," *Contemporary Music Review* 22, no. 4 (2003). (Note that in 2005 I changed my name from Caleb Stuart to Caleb Kelly.)

8. I should note here the convention I have used in relation to the order of Japanese names. Japanese names are given in traditional order: surname first. Exceptions are made for individuals who primarily reside outside of Japan and who have chosen to adopt the Western system: surname last.
9. Douglas Kahn, *Noise, Water, Meat* (Cambridge, Mass.: MIT Press, 1999), 101. A shorter version of this chapter has been printed as "The Sound of Music," in *The Auditory Culture Reader*, ed. Michael Bull and Les Back (Oxford: Berg, 2003), 77–90.
10. Douglas Kahn, *Noise, Water, Meat*, 102.
11. Ibid., 101.
12. Douglas Kahn, "Ether Ore: Mining Vibrations in American Modernist Music," in *Hearing Cultures: Essays on Sound, Listening, and Modernity*, ed. Veit Erlmann (Oxford: Berg, 2004), 112.
13. Ibid., 113.
14. The practices of extended and prepared techniques have been usefully discussed at length in Gary Butler's dissertation, "Prepared Instruments in Improvised Music: Precedents and Purposes" (Ph.D. diss., The University of Wollongong, 2000).

15. Jim Drobnick, "Listening Awry," in *Aural Cultures: Sound Art* (Banff: YYZ Books, 2004), 10, footnote 7. The term "sonic turn" is the audio version of W. J. T. Mitchell's "pictorial turn." See W. J. T. Mitchell, "The Pictorial Turn," *Artforum*, March 1992, 89–94.
16. Michael Bull and Les Back, "Introduction: Into Sound," in *The Auditory Culture Reader* (Oxford: Berg, 2003), 1.
17. Veit Erlmann, "But What of the Ethnographic Ear? Anthropology, Sound, and the Sense," *Hearing Cultures: Essays on Sound, Listening, and Modernity* (Oxford: Berg, 2004), 2.
18. Ibid., 3.
19. Christoph Cox and Daniel Warner, "Introduction: Music and the New Audio Culture," in *Audio Culture: Readings in Modern Music* (New York: Continuum, 2004), xiii.
20. Drobnick, "Listening Awry," 10.
21. Veit Erlmann, "But What of the Ethnographic Ear?," 2.
22. Bull and Back, "Introduction: Into Sound," 3.
23. Drobnick, "Listening Awry," 9.
24. Cox and Warner, "Introduction," xiii.
25. Kahn, *Noise, Water, Meat*, 2.
26. Jonathan Sterne, *The Audible Past: Cultural Origins of Sound Reproduction* (Durham: Duke University Press, 2003), 7.
27. Timothy Taylor, *Strange Sounds: Music, Technology, and Culture* (New York: Routledge, 2001), 25–26.

28. Mark Katz, *Capturing Sound: How Technology Has Changed Music* (Berkeley: University of California Press, 2004), 3.
29. Ibid., 3–4.
30. The term "software" is used here in the sense that the CD is the software to the CD player's hardware.
31. Frederick Kittler, *Gramophone, Film, and Typewriter* (Stanford: Stanford University Press, 1999), 1–2.
32. Sterne, *The Audible Past*, 283.
33. Ibid., 285.
34. Ibid.
35. See the section entitled "Destroyed Turntables at the End of Gramophonic Culture" in chapter 2 for a discussion of Otomo Yoshihide's work.
36. Media archaeology, as a methodology, is associated with a number of prominent academics including Friedrich Kittler, Erkki Hutamo, and Siegfried Zielinski.
37. Adorno, in a paper cowritten with Max Horkheimer entitled "The Culture Industry: Enlightenment as Mass Deception" (in Horkheimer and Adorno, *Dialectic of Enlightenment*, Stanford: Stanford University Press, 2002, 94–136), discusses the uniformity in various media, especially when they are brought together for film and television. This uniformity for economic convergence can be very clearly seen in the Apple iLife suite.
38. Lev Manovich, "Understanding Meta-Media," *Ctheory*, http://www.ctheory.net/articles.aspx?id=493 (accessed June 2008).

1 recording and noise: approaches to cracked media

1. Theodor Adorno, "The Curves of the Needle," *Essays on Music: Theodor Adorno* (Berkeley: University of California Press, 2002), 271.
2. Ibid.
3. A discussion of this line of thinking within Attali's text can be found in Jeff Schwartz, "Jacques Attali's Revision of Baudrillard in 'Noise,'" *Art-OMMA*, no. 9, 2003. http://www.art-omma.org/NEW/past_issues/theory/09.htm (accessed June 2008).
4. Bruce Andrews, "Praxis: A Political Economy of Noise and Informalism," *Close Listening: Poetry and the Performed Word*, ed. Charles Berstein (New York: Oxford University Press, 1998).
5. Evan Eisenberg, *The Recording Angel: The Experience of Music From Aristotle to Zappa* (New York: Penguin, 1987), 24.
6. Ibid., 25.
7. Walter Benjamin, "The Work of Art in the Age of Mechanical Reproduction," in *Illuminations* (London: Fontana, 1992), 214–215.
8. Ibid., 218.
9. Theodor Adorno, "The Radio Symphony: An Experiment in Theory," in *Essays on Music: Theodor Adorno* (Berkeley: University of California Press, 2002), 252.
10. Ibid., 256–261.
11. Michael Chanan, *Repeated Takes: A Short History of Recording and Its Effects on Music* (London: Verso, 1995), 9.

12. Theodor Adorno, "The Form of the Phonograph Record," in *Essays on Music: Theodor Adorno* (Berkeley: University of California Press, 2002), 278.
13. This is the key point of Attali's book: music prophesizes subsequent periods of history, foreshadowing capitalist developments in society.
14. Jacques Attali, *Noise: The Political Economy of Music* (Minneapolis: University of Minnesota Press, 1985), 88.
15. Ibid., 89.
16. Greg Hainge, "Come on Feel the Noise: Technology and Its Dysfunctions in the Music of Sensation," *To the Quick* 5 (2002), http://to-the-quick.binghamton.edu/pdf/tothequick5.pdf (accessed June 2008).
17. Ibid.
18. Ibid.
19. David Cunningham, "Goodbye 20th Century: Noise, Modernism, Aesthetics," transcript from the conference NoiseTheoryNoise 1, March 6, 2004.
20. Brandon LaBelle, "Music to the 'nth' Degree," in *Experimental Sound and Radio*, ed. Allen S. Weiss (Cambridge, Mass.: MIT Press, 2001), 167.
21. Cunningham, "Goodbye 20th Century."
22. Numerous academic papers have been written on Japanese noise music including: Paul Hegarty, "Full with Noise: Theory and Japanese Noise Music," *Ctheory*, http://www.ctheory.net/text_file.asp?pick=314 (accessed June 2008); Paul Hegarty, "Noise Threshold: Merzbow and the End of Natural Sound," *Organised Sound* 6, no. 3 (2001): 193–200; and Nick Smith, "The Splinter in

Your Ear: Noise as the Semblance of Critique," *Culture, Theory, and Critique* 46, no. 1 (2005): 43–59.

23. Smith, "The Splinter in Your Ear," 44.
24. Ibid.
25. Ibid.
26. For further discussion on Russolo in relation to experimental music and the avant-garde see Douglas Kahn, "Noises, Dada War Drums, and Russolo's Art of It," *Midwest* 8 (1995): 44–48.
27. Luigi Russolo, *The Art of Noises* (New York: Pendragon Press, 1986), http://www.ubu.com/papers/russolo.html (accessed June 2008).
28. Ibid.
29. Kahn, *Noise, Water, Meat*, 56–57.
30. Henry Cowell, *Essential Cowell: Selected Writings on Music* (Kingston, N.Y.: McPherson and Company, 2002), reprinted in *Audio Culture: Readings in Modern Music*, ed. Christopher Cox and Daniel Warner (New York: Continuum, 2004), 23–24.
31. John Cage, "The Future of Music: Credo," in *Silence* (London: Calder and Boyars, 1973, 1973), 3.
32. Herman Helmholtz, *On the Senesations of Tone* (New York: Dover, 1954), 7.
33. Ibid., 8.
34. Claude Shannon, "A Mathematical Theory of Communication," *Bell System Technical Journal* 27 (July and October 1948): 379–423, 623–656.
35. Claire Taylor, "Noise Is OK," *Semiotica* 52, no. 3/4 (1984): 274.

36. Paul Hegarty, *Noise/Music: A History* (New York: Continuum, 2007), 3.

37. Ibid., 199.

38. Michel Serres, *Genesis* (Michigan: University of Michigan Press, 1995), 7.

39. Michel Serres, *The Parasite* (Baltimore: Johns Hopkins University Press, 1982), 126.

40. Kahn, *Noise, Water, Meat*, 25.

41. Hegarty, "Full with Noise."

42. Attali, *Noise*, 3.

43. Ibid., 4.

44. Ibid., 26–27.

45. Ibid., 33.

2 broken music: the manipulated, modified, and destroyed phonograph

1. For simplicity, the term "phonograph" will be used henceforth to describe the phonograph proper as well as the gramophone. The phonograph is delimited by its use of a wax cylinder and also its ability to both record and playback sound; the gramophone differs from the phonograph in that it uses a flat disc, has to be wound up, and cannot record; and the "turntable" in modern usage describes any record player. "Turntable" will be specifically used here to describe the contemporary record player.

2. Charles Mudede, "The Turntable," *Ctheory*, http://ctheory.net/text_file.asp?pick=382 (accessed June 2008).
3. Ibid. Other texts that discuss the turntable as an instrument include: Bill Brewster and Frank Broughton, *Last Night a DJ Saved My Life* (London: Headline Book Publishing, 2000); Kodwo Eshun, *More Brilliant Than the Sun: Adventures in Sonic Fiction* (London: Quartet Books, 1998); Ulf Poschardt, *DJ Culture*, trans. Shaun Whiteside (London: Quartet Books, 1998); and Simon Reynolds, *Energy Flash* (London: Picador, 1998).
4. Gary Butler, "Prepared Instruments in Improvised Music: Precedents and Purposes" (Ph.D. diss., The University of Wollongong, 2000), 15.
5. The prepared instrument techniques developed by John Cage have been described in detail in other texts; see James Pritchett, *The Music of John Cage* (Cambridge: Cambridge University Press, 1993), 22–36; and David Revill, *The Roaring Silence: John Cage: A Life* (New York: Arcade Publishing, 1992), 69–71.
6. Chris Lawrence, "The Sonic Youth Gear Guide," http://www.sonicyouth.com/mustang/eq/gear.html (accessed July 2008).
7. See, for example, Mudede, "The Turntable."
8. Thomas Y. Levin, "Indexically Concretè: The Aesthetic Politics of Christian Marclay's Gramophonia," *Parkett* 56 (1999): 162.
9. Theodor Adorno, "The Curves of the Needle," reprinted in *Essays on Music: Theodor Adorno* (Berkeley: University of California Press, 2002), 275.
10. Benjamin, "The Work of Art in the Age of Mechanical Reproduction," 211–244.

11. Michael Chanan, *Repeated Takes: A Short History of Recording and Its Effects on Music* (London: Verso, 1995), 3.
12. Richard Schmidt James, "Expansion of Sound Resources in France, 1913–1940, and Its Relationship to Electronic Music" (Ph.D. diss., University of Michigan, 1981), and Mark Katz, "The Phonographic Effect: The Influence of Recording on Listener, Performer, Composer, 1900–1940" (Ph.D. diss., The University of Michigan, 1999).
13. Joel Chadabe, *Electric Sound: The Past and Promise of Electronic Music* (Englewood Cliffs, N.J.: Prentice-Hall, 1997), 28.
14. James, "Expansion of Sound Resources in France," 255–256.
15. Ibid., 258.
16. Raymond Lyon, "Le Phonograph d'avant-garde," *La Joie musicale* 3 (1930), quoted and translated by Richard Schmidt James, "Expansion of Sound Resources in France," 259.
17. Mark Katz, "The Phonographic Effect," 250–251.
18. Quoted in James, "Expansion of Sound Resources in France," 250.
19. James, "Expansion of Sound Resources in France," 260.
20. Mark Katz, "The Phonographic Effect," 227.
21. Evan Eisenberg, *The Recording Angel: Explorations in Phonography* (New York: McGraw-Hill, 1987), 129–130.
22. Katz, "The Phonographic Effect," 231.
23. Thomas Levin, "The Form of the Phonograph Record," *October* 55 (1990): 58, footnote 3.

24. See the section in chapter 3 entitled "Yasunao Tone's Wounded Compact Discs: From Improvisation and Indeterminate Composition to Glitching CDs."
25. László Moholy-Nagy, "New Plasticism in Music: Possibilities of the Gramophone," *Broken Music: Artists' Recordworks* (Berlin: DAAD-Gelbe Musik, 1989), 55–56.
26. Katz, "The Phonographic Effect," 241–242.
27. Cage, "The Future of Music: Credo," 4.
28. Ibid., 6.
29. Heinrich Burkhard, "Anmerkungen au den 'Lehrstücken' und zur Schallplattenmusik," *Melos* 9 (May/June 1930), quoted by Thomas Levin in "For the Record: Adorno on Music," *October* 55 (1990): 34.
30. Katz, "The Phonographic Effect," 234.
31. Ibid., 235–236.
32. Ernst Toch, "Über meine Kantate 'Das Wasswe' und meine Grammophonmusik," *Melos* 9 (May/June 1930), quoted in Katz, "The Phonographic Effect," 238.
33. Levin, "For the Record: Adorno on Music," 34.
34. Although Richard Schmidt James states that Varèse experimented with turntables around 1936, the later date of 1939 is more probably correct as it was during this period that he stopped composing to work on experiments for future directions in his composition.
35. Eisenberg, *The Recording Angel*, 133.

36. For an extended explanation of both the Philips Pavilion and Edgard Varèse's composition see Roberta Lukes, "The *Poème électronique* of Edgard Varèse" (Ph.D. diss., Harvard University, 1996).

37. John Cage, "Media Composition According to John Cage," in *Broken Music*, ed. Ursula Block and Michael Glasmeier (Berlin: DAAD-Gelbe Musik, 1989), 73.

38. Katz, "The Phonographic Effect," 256.

39. Ibid.

40. Ibid.

41. Virgil Thomson, "John Cage Late and Early (1960)," reprinted in *Writings about Cage*, ed. Richard Kostelanetz (Michigan: The University of Michigan Press, 1993), 103.

42. James Pritchett, *The Music of John Cage* (Cambridge: Cambridge University Press, 1993), 20.

43. John Cage, interview with Thom Holmes, 1981, in *Conversing with Cage*, ed. Richard Kostelanetz (New York: Limelight Editions, 1987), 158.

44. Revill, *The Roaring Silence*, 76.

45. Ibid., 198.

46. John Cage, quoted in *John Cage: An Anthology*, ed. Richard Kostelanetz (New York: Da Capo Press, 1991), 144.

47. For further discussion of the consequences of *Cartridge Music* see Richard Toop, "Writing Music's Boundaries," http://www.ssla.soc.usyd.edu.au/conference/toop.html (accessed June 2008).

48. Ken Friedman and James Lewes, "Fluxus: Global Community, Human Dimensions," *Visible Language* 26, no. 1/2 (1992): 156.
49. Douglas Kahn, "The Latest: Fluxus and Music," in *In the Spirit of Fluxus*, ed. Elizabeth Armstrong and Joan Rothfuss (Minneapolis: Walker Art Center, 1993), 102.
50. For an idea of how these artists relate to Fluxus, see Friedman and Lewes, "Fluxus: Global Community, Human Dimensions." The text includes a large table representing all of the artists involved in key Fluxus exhibitions, catalogs, and "Fluxlists."
51. David T. Doris, "Zen Vaudeville," in *Fluxus Reader*, ed. Ken Friedman (West Sussex: Academy Editions, 1999), 110.
52. Shiomi Mieko, quoted in Doris, "Zen Vaudeville," 110.
53. The word score for this piece is reproduced in Kahn's "The Latest," 105.
54. Shiomi Meiko, "Meiko Shiomi," *Art and Artists* (October 1973): 42.
55. Midori Yoshimoto, *Into Performance: Japanese Women Artists in New York* (New Brunswick: Rutgers University Press, 2005), 159.
56. Ibid., 164.
57. The word score for this work is reproduced in Kristine Stiles, "Between Water and Stone," in *In the Spirit of Fluxus*, 62.
58. The piece is known to a wider and younger audience as a result of the performances by Sonic Youth in their *Goodbye 20th Century* tour in Europe in 2000 and the documentation of their performance of Maciunas's work on the CD release *Goodbye 20th Century* (SYR, 1999).

59. Taehi Kang, "Nam June Paik: Early Years (1958–1973)" (Ph.D. diss., Florida State University, 1988), 68.
60. Ibid., 72.
61. Ina Blom, "Boredom and Oblivion," in *Fluxus Reader*, 77.
62. Kristine Stiles, "Thresholds of Control: Destruction Art and Terminal Culture," in *Ars Electronica: Facing the Future: A Survey of Two Decades*, ed. Timothy Druckrey (Cambridge, Mass.: MIT Press, 1999), 124.
63. Kristine Stiles, "The Destruction in Art Symposium (DIAS): The Radical Cultural Project of Event-Structured Live Art" (Ph.D. diss., University of California, Berkeley, 1987), 5–6.
64. Ibid., 6.
65. Ibid., 9.
66. For more information on the work of Gustav Metzger see Stiles, "The Destruction in Art Symposium (DIAS)"; Kristine Stiles, "Uncorrupted Joy: International Art Actions," in *Out of Actions: Between Performance and the Object, 1949–1979* (The Museum of Contemporary Art: Los Angeles, 1998); and Kristine Stiles, "Sticks and Stones: The Destruction in Art Symposium," *Arts Magazine* (January 1989): 54–60.
67. Stiles, "Sticks and Stones," 60, footnote 14.
68. Andrew Wilson, "Papa What Did You Do When the Nazis Built the Concentration Camps? My Dear, They Never Told Us Anything," in *Gustav Metzger: "Damaged Nature, Auto-Destructive Art"* (London: Coracle and Lanmet, 1996), 67–68.

69. Stiles, "The Destruction in Art Symposium (DIAS)," 396.
70. Ibid., 402–403.
71. Ibid., 403.
72. Kahn, "The Latest," 114.
73. Stiles, "The Destruction in Art Symposium (DIAS)," 406–407.
74. Milan Knízák, quoted in Stiles, "The Destruction in Art Symposium (DIAS)," 444.
75. Ibid., 444–445.
76. Ibid., 445.
77. Ibid., 446.
78. Martha Gever, "Pomp and Circumstances: The Coronation of Nam June Paik," *Afterimage* (October 1982): 12. Gever lists "his youth in Korea; his meeting with John Cage in Darmstadt, Germany; his gallery exhibition of altered TV sets in 1963" as examples.
79. John G. Harnhardt, *The Worlds of Nam June Paik* (New York: The Solomon R. Guggenheim Foundation, 2000), 20.
80. See Yong Woo Lee, "Nam June Paik's Teenage Years," in *Nam June Paik: eine DATA base*, ed. Klaus Bußman and Florian Matzner (Cologne: Edition Cantz, 1993) and Kang, "Nam June Paik: Early Years (1958–1973)," for a discussion of Paik's early carrier.
81. Calvin Tomkins, "Video Visionary," in *The Scene: Reports on Post-Modern Art* (New York: Viking Press, 1976), 200.
82. Harnhardt, *The Worlds of Nam June Paik*, 20.

83. Michael Nyman, "Nam June Paik, Composer," in *Nam June Paik*, ed. John G. Hanhardt (New York: Whitney Museum of Art in association with W. W. Norton, 1982), 82.

84. Kang, "Nam June Paik: Early Years," 34.

85. Harnhardt, *The Worlds of Nam June Paik*, 24.

86. Wulf Herzogenrath, "The Anti-technological Technology of Nam June Paik's Robots," in *Nam June Paik: Video Works 1963–88* (London: Heyward Gallery, 1988), 8.

87. Harnhardt, *The Worlds of Nam June Paik*, 35.

88. The work resonates with the much later work by Laurie Anderson; see section 5 below, "Christian Marclay and the Death of Vinyl."

89. Harry Ruhé, in *Fluxus: The Most Radical and Experimental Art Movement of the Sixties* (Amsterdam: A, 1979), states that Knízák organized the First Manifestation of Aktual art in 1964.

90. Milan Knížák, "Fluxus Debris," http://www.artnotart.com/fluxus/mknizak (accessed June 2008).

91. Milan Knížák, e-mail correspondence with author, October 6, 2001.

92. Milan Knížák, e-mail correspondence with author, August 24, 2001.

93. An image of this work appears in Ursula Block and Michael Glasmeier, eds., *Broken Music: Artists' Recordworks* (Berlin: DAAD-Gelbe Musik, 1989), 161.

94. Milan Knížák, *Broken Music* (CD, originally released by Multhipla Records, 1979, rereleased by Andromedia Records, 2002).

95. Milan Knížák, e-mail correspondence with author, October 6, 2001.

96. Reproduced in Block and Glasmeier, *Broken Music: Artists' Recordworks*, 161.
97. Milan Knížák, e-mail correspondence with author, October 6, 2001.
98. Ibid.
99. Christian Marclay, "Interview with Christian Marclay," by Jonathan Seliger, *Journal of Contemporary Art* 5 (spring 1992): 67.
100. "Interview with Christian Marclay," by Vincent Katz, *Print Collectors Newsletter* 22 (March/April 1991): 6.
101. Kahn, "The Latest," 109.
102. Marclay, interview by Katz, 6–7.
103. Marclay, interview by Seliger, 64.
104. Christian Marclay, "Interview with Christian Marclay," by Bice Curiger, *Züritip* 5 (October 1997): 55.
105. Marclay has played with many improvisers from New York and further afield, including: John Zorn, Nicolas Collins, Fred Frith, Haino Keiji, Bill Laswell, Arto Lindsay, Ikue Mori, Erik M, David Moss, Günter Müller, DJ Olive, Zeena Parkins, Jon Rose, Elliot Sharp, Sonic Youth, Yasunao Tone, and Otomo Yoshihide.
106. Kim Gordon, Matthew Higgs, and Jennifer Gonzalez, *Christian Marclay* (London: Phaidon, 2005) and Russell Ferguson, ed., *Christian Marclay* (London: Thames and Hudson, 2003).
107. Sonic Youth, *Goodbye 20th Century* (SYR, 1999).
108. Mark Dery, "Spin Doctors," *Keyboard* (October 1992): 58.
109. Marclay, interview by Seliger, 66.
110. Christian Marclay, liner notes to *Records (81–89)* (Atavistic, 1997).

111. For an interview exclusively about Marclay's early years see Douglas Kahn, "Christian Marclay's Early Years: An Interview," *Leonardo Music Journal* 13 (2003): 17–21.

112. Roselee Goldberg, *Laurie Anderson* (London: Thames and Hudson, 2000), 77.

113. Christian Marclay, conversation with Yasunao Tone, *Music* 1 (1997): 39.

114. Marclay, interview with Seliger, 66.

115. Gary Butler, "Prepared Instruments in Improvised Music," 234–241.

116. See Revill, *The Roaring Silence*, 187–190.

117. Christian Marclay, "Interview with Christian Marclay," by Jason Gross, *Perfect Sound Forever* (March 1998), http://www.furious.com/perfect/christianmarclay.html (accessed June 2008).

118. Poschardt, *DJ Culture*, 215–216.

119. Brewster and Broughton, *Last Night a DJ*, 278.

120. Marclay, interview by Gross.

121. Marclay, quoted in Mark Dery, "Spin Doctors," 59.

122. Thomas Brinkmann, quoted in Philip Sherburne, "Thumbnail Music: Six Artists Talk about Minimalism," *Urbansounds*, http://www.urbansounds.com/us_current/thumbnail/brinkmann_2.html (accessed April 1, 2000; no longer available online).

123. Joe Allen, "Backspinning Signifying," *To the Quick* 2 (1999), http://to-the-quick.binghamton.edu/issue%202/sampling.html (accessed June 2008). The article traces the use of breakbeats by hip-hop DJs. The breakbeat is a moment in a

tune where all the instruments except the drum stop for a couple of bars. These breaks are cut and repeated by the DJ during the set.

124. David Frankel, "Christian Marclay," catalog essay for *Christian Marclay 99.4*, International Artist-in-Residence program, ArtPace, San Antonio, Texas.
125. Ibid.
126. Marclay, interview by Gross.
127. Marclay, interview by Kahn, "Christian Marclay's Early Years," 19.
128. Marclay, conversation with Tone, 45.
129. Christian Marclay, "A Conversation between Christian Marclay and Michael Snow, New York, June 4, 2000," *Christian Marclay: Cinema*, exhibition held Dec. 9–Feb. 4, 2001, Oakville Galleries.
130. Christian Marclay, *Untitled*, 1987, 12-inch record and suede, edition of fifty, published by Ecart Editions.
131. Christian Marclay and Gary Hill, "Conversation," *Annandale* (spring 2000): 5.
132. There is little written on Gum. A retrospective anthology was released in 2004 entitled *Gum: Vinyl Anthology* (23five, 2004). The release includes liner notes by Jim Hayes.
133. Philip Samartzis, interview with author, January 19, 2006.
134. What Is Music? festival, held at Revolver, Melbourne, July 18, 2002. Video footage from this performance can be seen in episode 6 of *Subsonics* (SBS Television, Australia).
135. The exception is the Technics SL-1200, which is the turntable of choice within DJ culture.

136. Otomo Yoshihide's discography is large and goes some way to demonstrating the significance of his music. See *Improvised Music from Japan*, http://japanimprov.com/yotomo/disco.html (accessed June 2008).
137. Otomo Yoshihide, interview by Michel Henritz for *Revue et Corrigée*, October 2001. An English version can be found at http://www.japanimprov.com/yotomo/interview01.html (accessed June 2008).
138. See Atsushi Sasaki, "Dissolving Ground Zero: From 'Performance' to 'Sound,'" liner notes for Ground Zero, *Last Concert* (Amoebic, 1999).
139. Otomo, interview by Henritz.
140. Quoted in Clive Bell, "Music without Memory," *The Wire* 185 (July 1999): 36.
141. Otomo, interview by Henritz.
142. Sachiko M performs using a sampler that does not have any samples in its memory other than the test-tone that the device loads by default.
143. Hatanaka Minoru, "From an 'Off' Site," *Improvised Music from Japan* 1 (2002–2003): 8. This article gives a good overview of the gallery space and a description of the music created in it.
144. Butler, "Prepared Instruments in Improvised Music," 175.
145. Philip Samartzis, interview with the author, January 19, 2006.
146. Michael Graeve, artist's statement, e-mailed to author January 23, 2006.
147. Michael Graeve, e-mail interview with author, January 23, 2006.
148. Ibid.
149. Ibid.

150. Lucas Abela, "Interview with Lucas Abela," by Seymour Glass, *Bananafish* 12, posted at http://www.dualplover.com/Nananana.html (accessed May 2008).
151. Marclay, interview by Gross.

3 damaged sound: glitching and skipping compact discs in the audio of yasunao tone, nicolas collins, oval, and disc

A version of this chapter was originally published as "Damaged Sound: Glitching and Skipping Compact Discs in the Audio of Yasunao Tone, Nicolas Collins and Oval," *Leonardo Music Journal* 13 (2003): 47–52.

1. The occurrence of a skipping CD in a café may soon be a thing of the past as MP3 players, a more reliable playback medium, become more common.
2. Nicolas Collins, "Low Brass: The Evolution of Trombone-Propelled Electronics," *Leonardo Music Journal* 1, no. 1 (1991): 44.
3. David Ranada, "Understanding Compact Disc Errors," *Stereo Review* 49 (April 1984): 48. Ranada became *Stereo Review*'s champion of the CD system, writing a number of articles on the new format.
4. For a more detailed account see Ken Pohlmann, *The Compact Disc Handbook*, 2nd ed. (Madison, Wisconsin: A-R Editions, 1992).
5. Minor data losses, through scratches or dust, are covered or replaced by error-correction software that uses "redundant information" to fill in the gaps. Information is stored on the disc more than once, allowing for lost information

to be replaced and therefore have no effect on the audio. See Pohlmann, *The Compact Disc Handbook*, for more information on error correction and redundant information.

6. John Atkinson, "Compact Disc Again . . . John Atkinson Scratches His Head," *Hi-Fi News and Record Review* (November 1982): 29.
7. David Ranada, "Hands on the Digital Disc," *Stereo Review* 48 (April 1983): 63.
8. Ranada, "Understanding Compact Disc Errors," 48. For more detailed information on error-correcting algorithms and especially CIRC see Pohlmann, *The Compact Disc Handbook*, 60–74.
9. David Ranada, "Digital Debut: First Impressions of the Compact Disc System," *Stereo Review* 48 (December 1982): 70.
10. Ibid.
11. Ibid.
12. David Ranada, "How to Buy a CD Player," *Stereo Review* 48 (July 1983): 59. There is a photograph of the official test disc made by Philips next to Ranada's own creation. The Scotch tape example is of specific interest as it is what Yasunao Tone used when "wounding" his CDs.
13. A version of this section was originally published as "Yasunao Tone's Wounded and Skipping Compact Discs: From Improvisation and Indeterminate Composition to Glitching CDs," *Leonardo Electronic Almanac* 10, no. 9 (2002).
14. Alexandra Munroe, "A Box of Smile: Tokyo Fluxus, Conceptual Art, and the School of Metaphysics," in *Japanese Art after 1945: Scream against the Sky*, ed. Alexandra Munroe (New York: Harry N. Abrams, 1994), 215.

15. Alexandra Munroe, “The Spirit of YES: The Art and Life of Yoko Ono,” in *Yes Yoko Ono*, ed. Alexandra Munroe (New York: Harry N. Abrams, 2000), 19.
16. Ichiyanagi Toshi, “Essay by Toshi ICHIYANAGI,” *Japanese Art, 1960s*, http://www.arttowermito.or.jp/1960/music-e.html (accessed June 2008).
17. The list comes from Ken Friedman and James Lewes’s chart of artists who have been exhibited in twenty-one key Fluxus exhibitions. There are also several Japanese artists who have been in only one exhibition; these names are not found in the list above. See Friedman and Lewes, “Fluxus: Global Community, Human Dimensions,” *Visible Language* 26, no. 1/2 (1992): 155–179.
18. Yasunao Tone, from a transcript for an interview published in *Revue et Corrigée* 46 (2001).
19. The name Group Ongaku literally means “Group Music,” the paradox being that what they were creating might not be heard as music at all; that is, they were trying to produce “antimusic.” At the time, the Japanese avant-garde was waging a war against art, commercialism, and the Americanization of Japan. The artists involved in this protest took part in an institutional critique of the arts using forms of intermedia, happenings, and conceptualism, questioning the notions of art and creativity. Reiko Tomii, “Concerning the Institution of Art: Conceptualism in Japan,” in *Global Conceptualism: Points of Origin, 1950s–1980s* (New York: Queens Museum of Art, 1999), 17. See also Reiko Tomii, “State v. (Anti-)Art: Model 1,000-Yen Note Incident by Akasegawa Genpei and Company,” *Positions* 10, no. 1 (2002): 141–172.

20. An archival CD of Group Ongaku, entitled *Music of Group Ongaku* (Hear 002), was released in 1996. The CD includes two recordings of performances on May 8, 1960, at Mizuno's house, and September 15, 1961, at Sōgetsu Hall. Tojima Mikio, Tanno Yumiko, and Tsuge Genichi were also involved in Group Ongaku activities.
21. Tone, interview in *Revue et Corrigée*.
22. Yasunao Tone refers to the evening as *Improvisation and Objet Sonore* in his resume, but in the *Revue et Corrigée* interview he calls it *Concert of Sound Object and Improvisational Music.* It is important to note here that the word "objet" is a French-derived term and encompasses, in the Japanese art world, both "readymade and found objects" and "object-based work, such as combine, assemblage and junk art." See Tomii, "Concerning the Institution of Art," 19 and endnote 35.

    The term "Objet Sonore" was coined by Pierre Schaeffer; at the time Tone was influenced by Schaeffer's compositions and musique concrète.
23. Tone, quoted in Munroe, "A Box of Smile," 218.
24. Liz Kotz, "Post Cagean Aesthetics and the 'Event' Score," *October* 95 (2001): 57.
25. Both *Geodesy for Piano* and *Anagram for Strings* owe something to John Cage's scores for pieces such as *Cartridge Music* in their use of chance combinations and even the acetate sheet. Tone has stated in correspondence (July 21, 2001) that he knew little about Cage before Ichiyanagi's return to Tokyo late in 1961. His original understanding of chance had come from music journalism and

was very sketchy. The dates for both these pieces are 1962; therefore, they were written soon after Ichiyanagi returned and owe much to Tone's newfound knowledge of Cage and the New York School.

26. Tone, quoted in "On Tone," Christiane Kort, catalog essay for Loos Music Festival in The Hague, Netherlands, October 1999.
27. The recording of *Automatism* can be found on the CD *Music of Group Ongaku*.
28. Tone mentions this text in the liner notes to *Solo for Wounded CD* but no formal reference of this text has been found.
29. Yasunao Tone, liner notes to *Solo for Wounded CD* (Tzadik, 1997).
30. Tone, interview in *Revue et Corrigée*.
31. Tone, liner notes to *Solo for Wounded CD*.
32. Yasunao Tone, quoted in an interview with Christian Marclay, *Music* 1 (1997): 43.
33. John Cage, "Indeterminacy," in *Silence* (London: Calder and Boyars, 1973), 39.
34. Craig Kendal, liner notes to *Musica Iconologos* (Lovely Music, 1993).
35. Ibid.
36. Yasunao Tone, "John Cage and Record," *Intercommunication* (Japan) 35 (2000).
37. The title for the piece is derived from a work by Fluxus artist Alison Knowles entitled *Wounded Furniture*. The work, composed in 1965, is for "an old piece of furniture in bad shape. Destroy it further, if you like. Bandage it up with gauze and adhesive. Spray red paint on the wounded joints. Effective lighting helps. This activity may be performed with one or more performers, and simultaneously with other events." "Events by Alison Knowles," http://www.aknowles.com/eventscore.html (accessed June 2008).

38. Tone, correspondence with author, July 9, 2000.
39. Yasunao Tone, *Musica Simulacra* (Harvestworks Digital Arts Center, 1996).
40. Hannes Leopoldseder and Christine Schöpf, eds., *Cyberarts 2000* (Wein: Springer, 2000), 217.
41. See Nicolas Collins, "Cargo Cult Instruments," *Contemporary Music Review* 6 (1991): 73–84, for a description of early experimental musical instruments built by Collins.
42. Nicolas Collins, interview with John L. Walters, http://homestudio.thing.net/revue/content/collins2.htm (no longer available online).
43. Nicolas Collins, "Low Brass," 44.
44. Nicolas Collins, "In Conversation with Martin Conrads," http://www.art-bag.org/contd/issue2/collins.htm (accessed June 2008).
45. Ibid.
46. Ibid.
47. Collins, "Cargo Cult Instruments," 73.
48. Ibid.
49. Nicolas Collins, "Modified CD Players," http://members.nbci.com/_XMCM/nicollins/instruments.html#anchor1316096 (no longer available online).
50. Nicolas Collins, interview with John L. Walters. Collins has used the modified CD-players in a number of compositions and collaborations including: "Broken Light" on *It Was a Dark and Stormy Night* (Trace Elements, 1992), which uses a performer-controlled skipping CD, as does "Still Lives" (for trumpet, voice, and skipping CD) on *Sound without Picture* (Periplum, 1999). *A Host, of*

Golden Daffodils (Periplum, 1999) is made up of "improvised duos with Peter Cusack" and is "strewn with CD artifacts." Information from author's e-mail correspondence with Nicolas Collins, November 16, 2000.

51. Nicolas Collins, e-mail correspondence with the author, November 27, 2000.
52. Collins has used CD technology in group situations and other projects. He has worked with the group Impossible Music, formed by David Weinstein and Tim Spelios, which included Ted Greenwald, Ikue Mori, and David Shea. Collins also produced a CD entitled *100 of the World's Most Beautiful Melodies* (Trace Elements Records, 1988), which made use of the random access button on the CD player. Collins believes that this CD is the first to deliberately use the random function button as a compositional playback tool.
53. Simon Reynolds, *Energy Flash* (London: Picador, 1998), 389.
54. Kim Cascone, "Popp Art," *Artbyte* 3, no. 4 (November/December 2000): 77.
55. Frank Metzger, interviewer unknown, April/May 1999, http://the.meso.net/oschatz/oval/metzger.html (June 3, 2008). In its first incarnation Oval also included Holger Lindmüller.
56. Frank Metzger, e-mail correspondence with the author, January 29, 2001.
57. Sebastian Oschatz, e-mail correspondence with the author, January 17, 2001.
58. Sebastian Oschatz, e-mail correspondence with the author, September 12, 2000. It is interesting that there was some controversy over some of the sources for the skips and stutters on the later album *Systemisch*. It was claimed that many of the sounds were sourced from an album by British producer Aphex

Twin. This was confirmed in an interview with Metzger and also by Oschatz in my correspondence.

59. Sebastian Oschatz, e-mail correspondence with author, January 17, 2001.
60. Jörg Heiser, "Dechingdechingdeching," *Heaven Sent* (February 1994). Translation from the original German.
61. Ibid.
62. Sascha Kölsch, "Oval Aners Als Die Zeit," *Spex* (January 1994): 26–27. Translation from the original German.
63. Sebastian Oschatz, e-mail correspondence with the author, January 17, 2001.
64. Oval, *Wohnton Installation*, http://the.meso.net/oschatz/oval/wohntoninst.html (accessed June 2008).
65. Mille Plateaux developed a name for releasing interesting and experimental new music. For a critical overview see Simon Reynolds, "Low End Theories," *The Wire* 146 (April 1996). Mille Plateaux also signed over the release to American independent label Thrill Jockey for a rerelease in 1996. Note: many articles and papers date the *Systemisch* release as 1996, as the Thrill Jockey reissue has this date on the cover of the CD package.
66. The lack of lyrics is in stark contrast to the vocals found in Oval's earlier work. The move away from the style was forced by the need to have a freer space for the new samples and sequences to play out. The text and vocals were heard as too restrictive in structure for the new direction. Moreover, at the time of its production of *Systemisch* Oval did not have access to CD-burning technology,

and therefore Metzger could not simply record his vocals to CD-R and then affect the disc for sampling.

67. Frank Metzger, e-mail correspondence with the author, September 26, 2001.
68. Markus Popp, quoted in A. C. Lee, "POPP ART: An Interface with Oval," *Tuba Frenzy* 4 (May 1998).
69. Lev Manovich, *The Language of New Media* (Cambridge, Mass.: MIT Press, 2001), 46.
70. The *Ovalprocess* project was begun in 1995–1996 and in its final version was a collaboration between Popp and Richard Ross. The work was originally conceived while Oschatz was part of the group. In its installation form the work has been shown at a number of galleries in Europe and and the United States since 2001, and it received an honorary mention in the 2001 Ars Electronica awards.
71. Some of the tracks Oval had produced in 1994 and not included on *Systemisch* appeared on the later release *94diskont.*
72. Popp makes numerous references to this in a number of interviews. See Reynolds, "Low End Theories" and Lee, "POPP ART."
73. Popp, subsequent to the release of *94discont*, became the sole artist working under the moniker of Oval. Oschatz and Metzger were removed in circumstances that fit the stereotype of a rock band breakup rather than a group involved in the digital audio scene. The tension this has created makes it hard to draw primary answers from Popp himself, and as such the previous sections

have relied mainly on discussions with Metzger and Oschatz, with cross references between their respective answers and printed media on Oval.

74. Paul Schütze, "Review of Oval, *94diskont*" (Mille Plateaux, MP13), *The Wire* 135 (May 1995): 48.
75. Simon Reynolds, *Energy Flash* (London: Picador, 1998), 390.
76. Nabeel Zuberi, *Sounds English: Transnational Popular Music* (Urbana: Illinois University Press 2001), 147.
77. Paul Théberge, *Any Sound You Can Imagine: Making Music/Consuming Technology* (Middleton, Conn.: Wesleyan University Press, 1997), 5–6.
78. Ibid., 243.
79. Quoted in Reynolds, "Low End Theories." Oval limited its use of music production tools to commercial music technologies. This is one response to the problem of determinism within the market; the other is to create new tools yourself. This approach can be linked back to Nicolas Collins and in a wider sense to the practices of instrument building, such as those who utilize MAX/MSP and Pure Data, within digital environments.
80. Cascone, "Popp Art," 77.
81. Author unknown, interview with J Lesser, 1998, http://www.kdvs.org/kdviations/summer98/lsr.html (no longer available online).
82. Ibid.
83. J Lesser, e-mail interview with the author, July 23, 2001.

4 tactics, shadows, and new media

1. Guy Debord, *The Society of the Spectacle* (Detroit: Black and Red, 1983); Jean Baudrillard, *Simulations* (New York: Semiotext[e], 1983); Michel de Certeau, *The Practice of Everyday Life* (Berkeley: University of California Press, 1988); Pierre Bourdieu, *The Field of Cultural Production: Essays on Art and Literature* (Cambridge: Polity Press, 1993); Nicolas Bourriaud, *Postproduction: Culture as Screenplay* (New York: Lucas and Sternberg, 2002).
2. Hakim Bey, *T.A.Z.: The Temporary Autonomous Zone, Ontological Anarchy, Poetic Terrorism* (New York: Autonomedia, 1985).
3. See the Reclaim the Streets website, http://rts.gn.apc.org/ (accessed June 2008).
4. Graffiti Research Lab, "LED Throwies," http://graffitiresearchlab.com/?page_id=6#video (accessed March 2008).
5. De Certeau, *The Practice of Everyday Life*, xi.
6. Ibid.
7. Ibid., 36.
8. Ibid., 37.
9. Ibid.
10. Ibid., xix.
11. Attali, *Noise*, 134.
12. Ibid.

13. Simon Grant, "Ceal Floyer," in *Everyday: 11th Biennale of Sydney* (Sydney: Biennale of Sydney, 1998), 96.
14. Simon Grant, "Clay Ketter," in *Everyday: 11th Biennale of Sydney*, 128.
15. See Nicolas Bourriaud, *Relational Aesthetics* (Paris: Les Presses du Réel, 1998).
16. Nikos Papastergiadis, "'Everything That Surrounds': Art, Politics, and Theories of the Everyday," in *Everyday: 11th Biennale of Sydney*, 23.
17. Paul DeMarinis, interview by Shun-ichi Shiba, http://www.well.com/user/demarini/shiba.html (accessed June 2008).
18. Paul DeMarinis, Artist's talk delivered at International Conference on Auditory Displays, Palo Alto, November, 1997, http://www.well.com/~demarini/icad.html (accessed June 2008).
19. Ibid.
20. Art Lange, "Martin Tétreault: Broken Music Maker," *The Wire* 180 (February 1999): 12–13.
21. Kim Cascone, press release for *Pulsar Studies 1-20*, November 15, 1999.
22. Kim Cascone, "Residualism," http://www.anechoicmedia.com/residualism/ (no-longer available on-line).
23. Ibid.
24. Kim Cascone, "The Aesthetics of Failure: 'Post-Digital' Tendencies in Contemporary Computer Music," *Computer Music Journal* 24, no. 4 (2000): 13–14.
25. Ibid., 13.

26. The idea for the work is reminiscent of American experimental composer Earle Brown's work *Octet* (1952–53), which used the off-cuts from John Cage's *William's Mix* (1952).
27. The site http://fals.ch is no longer available online.
28. Sterne, *The Audible Past*, 283.
29. Mitchell Whitelaw, "Inframedia Audio," *Artlink* 21, no. 3 (2001): 51.
30. Manovich, *The Language of New Media*, 36.
31. Ibid., 41.
32. Ibid., 42.
33. Jeroen de Boer, "Radian, May 22," e-mail sent to various e-mail lists, May 9, 2000.
34. Tim Owen, "Review of *Faßt* (Touch, TO:32)," *The Wire* 162 (August 1997): 52.
35. See Tina Frank's website for examples of her practice: http://www.frank.at (accessed June 2008).
36. Examples of websites produced by Jodi include http://jodi.org and http://404.jodi.org (accessed June 2008).
37. Margaret Sundall, "Jodi—New York—Internet art of Joan Heemskerk and Dirk Paesmans featured in INSTALL.EXE," *Art Forum* (September 2003).
38. The Fairlight CVI is an analog visual synthesizer released on the market in 1984 and designed by Australians Peter Vogel and Kim Ryrie.

bibliography

Abela, L. "Interview with Lucas Abela." By S. Glass. *Bananafish* 12. Available at http://www.dualplover.com/Nanananana.html.

"A Conversation between Christian Marclay and Michael Snow, New York, June 4, 2000." In *Christian Marclay: Cinema*, ed. B. Portis. (Exhibition held December 2000–February 2001.) Oakville, Ontario: Oakville Galleries, 2000.

Adorno, T. *Essays on Music: Theodor Adorno*. Ed. Richard Leppert. Berkeley: University of California Press, 2002.

Allen, J. "Backspinning Signifying." *To the Quick* 2 (1999). http://to-the-quick.binghamton.edu/issue%202/sampling.html.

Anderson, C. "Cat Scratch Fever." *EQ Magazine* (August 1993): 114–115.

Andrews, B. "Praxis: A Polical Economy of Noise and Informalism." In *Close Listening: Poetry and the Performed Word*, ed. Charles Bernstein. New York: Oxford University Press, 1998.

Atkinson, J. "Compact Disc Again . . . John Atkinson Scratches His Head." *Hi-Fi News and Record Review* (November 1982).

Attali, J. *Noise: The Political Economy of Music*. Minneapolis: University of Minnesota Press, 2002.

Auslander, P. "Looking at Records." In *Aural Cultures: Sound Art*, ed. Jim Drobnick, 150–156. Banff: YYZ Books, 2004.

Baert, L., ed. *Digital Audio and Compact Disc Technology*, 3rd ed. Oxford: Focal Press, 1995.

Baudrillard, J. *Simulations.* New York: Semiotext(e), 1983.

Baxter, E. "Normal Japanese Child: Otomo Interviewed by Ed Baxter." *Resonance* 4:2 (1996): 8–11.

Bell, C. "Music without Memory." *The Wire* 185 (July 1999): 32–36.

Benjamin, W. "Art in the Age of Mechanical Reproduction." In *Illuminations*, 211–244. London: Fontana, 1992.

Bey, H. *T.A.Z.: The Temporary Autonomous Zone, Ontological Anarchy, Poetic Terrorism.* New York: Autonomedia, 1985.

Blake, A., ed. *Living through Pop*. London: Routledge, 1999.

Block, U., and M. Glasmeier, eds. *Broken Music: Artists' Recordworks*. Berlin: DAAD-Gelbe Musik, 1989.

Blom, I. "Boredom and Oblivion." In *Fluxus Reader*, ed. Ken Friedman, 64–90. West Sussex: Academy Editions, 1999.

Born, G. "Modern Music Culture: On Shock, Pop, and Synthesis." *New Formations* 2 (1987): 51–78.

Borwick, J. "The New Sound System: A Beginner's Guide." *Gramophone* (March 1983): 1001–1002.

Bourdieu, P. *The Field of Cultural Production: Essays on Art and Literature.* Cambridge: Polity Press, 1993.

Bourriaud, N. *Postproduction: Culture as Screenplay.* New York: Lucas and Sternberg, 2002.

Bourriaud, N. *Relational Aesthetics.* Paris: Les Presses du Réel, 1998.
Brewster, B., and F. Broughton. *Last Night a DJ Saved My Life*. London: Headline Book Publishing, 2000.
Bull, M., and L. Back. “Introduction: Into Sound.” In *The Auditory Culture Reader*, ed. Michael Bull and Les Back, 1–23. Oxford: Berg, 2003.
Burkhard, H. “Anmerkungen au den ‘Lehrstücken’ und zur Schallplattenmusik.” *Melos* 9 (May/June 1930).
Butler, G. “Prepared Instruments in Improvised Music: Precedents and Purposes.” Ph.D. diss., University of Wollongong, 2000.
Cadet, N. “Christian Marclay.” *High Performance* 10:38 (1987): 73–74.
Cage, J. “The Future of Music: Credo.” In *Silence.* London: Calder and Boyars, 1973.
Cage, J. *Silence*. London: Calder and Boyars, 1973.
Cascone, K. “Popp Art.” *Artbyte* 3, no. 4 (2000): 76–78.
Cascone, K. Press release for *Pulsar Studies 1–20* (15 November 1999).
Cascone, K. “Residualism.” http://www.anechoicmedia.com/residualism/silence.html (no longer available online).
Cascone, K. “The Aesthetics of Failure: ‘Post-Digital’ Tendencies in Contemporary Computer Music.” *Computer Music Journal* 24, no. 4 (2000): 12–18.
Celant, G. *The Record as Artwork: From Futurism to Conceptual Art*. Fort Worth, Texas: The Fort Worth Art Museum, 1977.
Chadabe, J. *Electric Sound: The Past and Promise of Electronic Music*. Englewood Cliffs, N.J.: Prentice-Hall, 1997.

Chanan, M. *Repeated Takes: A Short History of Recording and Its Effects on Music*. London: Verso, 1995.

Collins, N. "All This and Brains Too: Thirty Years of Howling Round." *Resonance* 9:2 (2002).

Collins, N. "Cargo Cult Instruments." *Contemporary Music Review* 6 (1991): 73–84.

Collins, N. "In Conversation with Martin Conrads." http://www.art-bag.org/contd/issue2/collins.htm.

Collins, N. "Interview with Nicolas Collins." By John L. Walters. http://homestudio.thing.net/revue/content/collins2.htm (no-longer available on-line).

Collins, N. "Low Brass: The Evolution of Trombone-Propelled Electronics." *Leonardo Music Journal* 1, no. 1 (1991): 41–44.

Collins, N. "Ubiquitous Electronics—Technology and Live Performance 1966–1996." *Leonardo Music Journal* 8 (1998): 27–32.

Conversation between Christian Marclay and Yasunao Tone. *Music*, no. 1, 1997.

Cott, J. *Stockhausen: Conversions with the Composer*. New York: Simon and Schuster, 1973.

Coulter, T. "Viva Vinyl: The Record as Art." *Ear* 15:3 (1990): 16–19.

Cowell, H. "The Joys of Noise." In *Audio Culture: Readings in Modern Music*, ed. Christoph Cox and Daniel Warner, 22–24. New York: Continuum, 2004.

Cox, C., and D. Warner, eds. *Audio Culture: Readings in Modern Music.* New York: Continuum, 2004.

Cox, C., and D. Warner. "Introduction: Music and the New Audio Culture." In *Audio Culture: Readings in Modern Music*, ed. Christoph Cox and Daniel Warner, xiii–xvii. New York: Continuum, 2004.

Culter, C. "A History of Plunderphonics." *Resonance* 3, no. 2 (1995).

Cunningham, D. "Goodbye 20th Century: Noise, Modernism, Aesthetics." Transcript from *NoiseTheoryNoise 1* (March 6, 2004).

Debord, G. *The Society of the Spectacle*. Detroit: Black and Red, 1983.

de Certeau, M. *The Practice of Everyday Life*. Berkeley: University of California Press, 1988.

Decker-Phillips, E. *Paik Video*. Barrytown: Station Hill Arts, 1997.

DeMarinis, P. Artist's talk delivered at International Conference on Auditory Displays, Palo Alto, November 1997. http://www.well.com/~demarini/icad.html.

DeMarinis, P. Interview by Shun-ichi Shiba. Originally published 1997 (in Japanese) in *ICC Journal*. English translation available at http://www.well.com/user/demarini/shiba.html.

Derschmidt, E. "The Disappearance of the 'Jazu-Kissa': Some Considerations about Japanese 'Jazz-Café's' and Jazz-Listeners." In *The Culture of Japan as Seen Through Its Leisure,* ed. Sepp Linhardt and Sabibe Frühstück, 303–315. New York: State University of New York Press, 1998.

Dery, M. "Spin Doctors." *Keyboard* (October 1992): 54–69.

Dery, M. "The Culture Jammers." *Keyboard* (August 1990): 57–70.

Diederichsen, D. Review of *Oval Wohton*. *Spex* (November 1993).

Doris, D. T. “Zen Vaudeville.” In *Fluxus Reader*, ed. Ken Friedman, 91–135. West Sussex: Academy Editions, 1999.

Drobnick, J. “Listening Awry.” In *Aural Cultures: Sound Art*, ed. Jim Drobnick, 9–15. Drobnick. Banff: YYZ Books, 2004.

Duguid, B. Review of *Ovalprocess*. *The Wire* 196 (June 2000): 50.

Eisenberg, E. *The Recording Angel: The Experience of Music From Aristotle to Zappa*. New York: Penguin, 1987.

Elliot, D., ed. *Reconstructions: Avant-Garde Art in Japan 1945–1965*. Oxford: Museum of Modern Art (Oxford) and Universe Books, 1985.

Erlmann, V. “But What of the Ethnographic Ear? Anthropology, Sound, and the Sense.” In *Hearing Cultures: Essays on Sound, Listening and Modernity*, ed. Veit Erlmann, 1–20. Oxford: Berg, 2004.

Eshun, K. *More Brilliant Than the Sun: Adventures in Sonic Fiction*. London: Quartet Books, 1998.

Ferguson, R., ed. *Christian Marclay*. London: Thames and Hudson, 2003.

Frankel, D. “Christian Marclay.” Catalog essay for *Christian Marclay 99.4*. International Artist-in-Residence program, ArtPace, San Antonio, Texas.

Friedman, K., and J. Lewes. “Fluxus: Global Community, Human Dimensions.” *Visible Language* 26, no. 1/2 (1992): 155–179.

Frith, S., and H. Horne. *Art Into Pop*. Methuen: London, 1987.

Gann, K. Review of *Solo for Wounded CD*. *Village Voice* 36 (September 9, 1997).

Gever, M. “Pomp and Circumstances: The Coronation of Nam June Paik.” *Afterimage* (October 1982): 12–16.

Gitelman, L., and G. Pingree, eds. *New Media, 1740–1915*. Cambridge, Mass.: MIT Press, 2003.

Gold, A. "Hands-on CD with Alvin Gold." *Hi-Fi News and Record Review* (March 1983).

Goldberg, R. *Laurie Anderson*. London: Thames and Hudson, 2000.

Gordon, K., M. Higgs, and J. Gonzalez. *Christian Marclay*. London: Phaidon, 2005.

Grant, G. "Ceal Floyer." In *Everyday: 11th Biennale of Sydney.* Sydney: Biennale of Sydney, 1998.

Grant, G. "Clay Ketter." In *Everyday: 11th Biennale of Sydney.* Sydney: Biennale of Sydney, 1998.

Greenberg, C. "Avant-Garde and Kitsch." In *Art and Culture*, 3–21. London: Thames and Hudson, 1973.

Hainge, G. "Come on Feel the Noise: Technology and Its Dysfunctions in the Music of Sensation." *To the Quick* 5 (2002). http://to-the-quick.binghamton.edu/pdf/tothequick5.pdf.

Hainge, G. "On the Very Possibility of Noise Music." Unpublished paper presented at CongressCATH, Leeds, 2003.

Hansen, M. *New Philosophy for New Media*. Cambridge, Mass.: MIT Press, 2004.

Harnhardt, J. G. *Nam June Paik*. New York: Whitney Museum of American Art, 1982.

Harnhardt, J. G. *The Worlds of Nam June Paik*. New York: The Solomon R. Guggenheim Foundation, 2000.

Hatanaka, M. "From an 'Off' Site." *Improvised Music from Japan* 1 (2002–2003).

Hegarty, P. "Full with Noise: Theory and Japanese Noise Music." *Ctheory*. http://www.ctheory.net/text_file.asp?pick=314.

Hegarty, P. *Noise/Music: A History*. New York: Continuum, 2007.

Hegarty, P. "Noise Threshold: Merzbow and the End of Natural Sound." *Organised Sound* 6, no. 3 (2001): 193–200.

Heiser, J. "Dechingdechingdeching." *Heaven Sent* (February 1994).

Helmholtz, H. *On the Sensations of Tone*. New York: Dover, 1954.

Hendricks, J. "Yoko Ono and Fluxus." In *Yes Yoko Ono*, ed. Alexandra Monroe, 38–50. New York: Harry N. Abrams, 2000.

Herzogenrath, W. "The Anti-Technological Technology of Nam June Paik's Robots." In *Nam June Paik: Video Works 1963–88*, 6–31. London: Hayward Gallery, 1988.

Higgins, D. "Intermedia." *Leonardo Music Journal* 34: 1 (2001): 49–54.

Higgins, D. "Synesthesia and Intersenses: Intermedia." *Something Else Newsletter* 1 (1966). http://www.ubu.com/papers/higgins_intermedia.html.

Higgins, H. "Fluxus Fortuna." In *Fluxus Reader*, ed. Ken Friedman, 31–60. West Sussex: Academy Editions, 1999.

Horkheimer, M., and T. Adorno. *Dialectic of Enlightenment*. Stanford: Stanford University Press, 2002.

Ichiyanagi, T. "Essay by Toshi ICHIYANAGI." *Japanese Art, 1960s.* http://www.arttowermito.or.jp/1960/music-e.html.

Ippolito, I. "Ten Myths of Internet Art." *Leonardo* 35:5 (2002): 485–498.

James, R. S. "Expansion of Sound Resources in France, 1913–1940, and Its Relationship to Electronic Music." Ph.D. diss., University of Michigan, 1981.

Jerkins, J., ed. *In the Spirit of Fluxus*. Minneapolis: Walker Art Center, 1993.

Kahn, D. "Audio Art in the Deaf Century." In *Sound by Artists*, ed. Dan Lander and Micah Lexier, 301–328. Toronto: Art Metroploe, 1990.

Kahn, D. "Christian Marclay Is Not a Sound Artist." Museum of Contemporary Art, Chicago. http://www.mcachicago.org/cm_media/run-index.htm (no longer available online).

Kahn, D. "Christian Marclay's Early Years: An Interview." *Leonardo Music Journal* 13 (2003): 17–21.

Kahn, D. "Ether Ore: Mining Vibrations in American Modernist Music." In *Hearing Cultures: Essays on Sound, Listening, and Modernity*, ed. Veit Erlmann, 107–130. Oxford: Berg, 2004.

Kahn, D. "Introduction: Histories of Sound Once Removed." In *Wireless Imagination: Sound, Radio, and the Avant-Garde*, 1–29. Cambridge, Mass.: MIT Press, 1994.

Kahn, D. *Noise, Water, Meat: A History of Sound in the Arts*. Cambridge, Mass.: MIT Press, 1999.

Kahn, D. "Noises, Dada War Drums, and Russolo's Art of It." *Midwest*, no. 8 (1995): 44–48.

Kahn, D. "The Latest: Fluxus and Music." In *In the Spirit of Fluxus*, ed. J. Jenkins, 100–121. Minneapolis: Walker Art Center, 1993.

Kahn, D. "The Sound of Music." In *The Auditory Culture Reader*, ed. Michael Bull and Les Back, 77–90. Oxford: Berg, 2003.

Kahn, D. "Track Organology." *October* 55 (1990): 67–78.

Kang, T. "Nam June Paik: Early Years (1958–1973)." Ph.D. diss., Florida Sate University, 1988.

Katz, M. *Capturing Sound: How Technology Has Changed Music*. Berkeley: University of California Press, 2004.

Katz, M. "The Phonographic Effect: The Influence of Recording on Listener, Performer, Composer, 1900–1940." Ph.D. diss., The University of Michigan, 1999.

Kendall, C. Liner notes to *Musica Iconologos*. Lovely Music, 1993.

Khazam, R. "Erik M: Sticky Fingers." *The Wire* 196 (June 2000): 11–12.

Khazam, R. "Jumpcut Jockey." *The Wire* 195 (May 2000): 28–33.

Kittler, F. *Gramophone, Film, and Typewriter*. Stanford: Stanford University Press, 1999.

Knížák, M. "Broken Music." In *Broken Music: Artists' Recordworks*, ed. Ursula Block and Michael Glasmeier. Berlin: DAAD-Gelbe Musik, 1989.

Knížák, M. *Broken Music*. CD, originally released by Multhipla Records, 1979. Rereleased by Andromedia Records, 2002.

Knížák, M. "Fluxus Debris." http://www.artnotart.com/fluxus/mknizak.

Kölsch, S. "Oval Aners Als Die Zeit." *Spex* (January 1994).

Kort, C. "On Tone." Catalog essay for Loos Music Festival in The Hague, Netherlands, October 1999.

Kostelanetz, R., ed. *Conversing with Cage*. New York: Limelight Editions, 1987.

Kostelanetz, R., ed. *John Cage: An Anthology*. New York: Da Capo Press, 1991.

Kostelanetz, R., ed. *Writings about Cage*. Michigan: The University of Michigan Press, 1993.

Kosugi, T. "Fishing for Sound." Interview with John Hudak. In *Interspersions*. Berlin: DAAD Gallery, 1992.

Kotz, L. "Post Cagean Aesthetics and the 'Event' Score." *October* 95 (2001): 55–89.

Krasnow, D. "Spin Doctor." *Artforum* (November 1998): 43.

Krauss, R. "Video: The Aesthetics of Narcissism." *October* 1 (spring 1976): 50–64.

LaBelle, B. "Music to the 'nth' Degree." In *Experimental Sound and Radio*, ed. Allen S. Weiss, 167–174. Cambridge, Mass.: MIT Press, 2000.

Lange, A. "Martin Tétreault: Broken Music Maker." *The Wire* 180 (February 1999): 12–13.

Latartara, J. "Merzbow, Spiral Blast, and the Ecstasy of Sound Itself." *SAPAAN* 1 (2003). http://www.sapaan.com/vol1/latartar.htm.

Lee, A. C. "POPP ART: An Interface with Oval." *Tubafrenzy* 4 (May 1998).

Lee, Y. W. "Nam June Paik's Teenage Years." In *Nam June Paik: eine DATA base*, ed. Klaus Bußman and Florian Matzner, 22–26. Cologne: Edition Cantz, 1993.

Leopoldseder, H., and C. Schöpf, eds. *Cyberarts 2000*. Wein: Springer, 2000.

Lesser, J (author unknown). Interview (1998). http://www.kdvs.org/kdviations/summer98/lsr.html (no longer available online).

Levin, T. "For the Record: Adorno on Music in the Age of Its Technological Reproducibility." *October* 55 (1990): 23–47.

Levin, T. "Indexically Concretè: The Aesthetic Politics of Christian Marclay's Gramophonia." *Parkett* 56 (1999): 162–175.

Loubet, E. "Laptop Performers. Compact Disc Designers, and No-Beat Techno Artists in Japan: Music from Nowhere." *Computer Music Journal* 24:4 (2000): 19–32.

Loubet, E. "The Beginnings of Electronic Music in Japan, with a Focus on the NHK Studio: The 1950s and 1960s." *Computer Music Journal* 21, no. 4 (1997): 11–22.

Loubet, E. "The Beginnings of Electronic Music in Japan, with a Focus on the NHK Studio: The 1970s." *Computer Music Journal* 22, no. 1 (1998): 49–55.

Lukes, R. "The *Poème électronique* of Edgard Varèse." Ph.D. diss., Harvard University, 1996.

Lyon, R. "Le Phonograph d'avant-garde." *La Joie musicale* 3 (1930).

Maconie, R. *The Works of Karlheinz Stockhausen*. London: Oxford University Press, 1976.

Manovich, L. *The Language of New Media*. Cambridge, Mass.: MIT Press, 2001.

Manovich, L. "Understanding Meta-Media." *Ctheory*, http://www.ctheory.net/articles.aspx?id=493 (3 June 2008).

Marclay, C. "Interview with Christian Marclay." By B. Curiger. *Züritip* 5 (October 1997): 55–61.

Marclay, C. "Interview with Christian Marclay." By J. Gross. *Perfect Sound Forever* (March 1998). http://www.furious.com/perfect/christianmarclay.html.

Marclay, C. "Interview with Christian Marclay." By V. Katz. *Print Collectors Newsletter* 22 (March/April 1991): 5–9.

Marclay, C. "Interview with Christian Marclay." By J. Seliger. *Journal of Contemporary Art* 5, no. 1 (1992): 64–76.

Marclay, C. "Conversation with Yasunao Tone." *Music* 1 (1997).

Marclay, C., and G. Hill. "Conversation." *Annandale* (spring 2000): 2–9.

McClary, S. "Terminal Prestige: The Case of Avant-Garde Music Composition." In *Keeping Score: Music, Disciplinary, Culture*, ed. David Schwarz, Anahid Kassabian, and Lawrence Siegel, 27–46. Charlottesville, Virginia: University Press of Virginia, 1997.

Metzger, F. Interview April/May 1999. http://the.meso.net/oschatz/oval/metzger.html.

Minoru, H. "From an 'Off' Site." *Improvised Music from Japan* 1 (2002–2003): 8.

Mitchell, W. J. T. "The Pictorial Turn." *Artforum* (March 1992): 89–94.

Moholoy-Nagy, L. "New Plasticism in Music: Possibilities of the Gramophone." In *Broken Music: Artists' Recordworks*, 55–56. Berlin: Daadgalerie, 1989.

Moore, T. Liner notes for Christian Marclay, *Records*. CD, Atavistic, 1997.

Motherwell, R. *The Dada Painters and Poets: An Anthology*. Cambridge, Mass.: Harvard University Press, 1981 (first ed., 1951).

Mudede, C. "The Turntable." *Ctheory*. http://ctheory.net/text_file.asp?pick=382.

Munroe, A. "A Box of Smile: Tokyo Fluxus, Conceptual Art, and the School of Metaphysics." In *Japanese Art after 1945: Scream against the Sky*, ed. Alexandra Munroe, 215–225. New York: Harry N. Abrams, 1994.

Munroe, A., ed. *Japanese Art after 1945: Scream against the Sky*. New York: Harry N. Abrams, 1994.

Munroe, A. "The Spirit of YES: The Art and Life of Yoko Ono." In *Yes Yoko Ono*, ed. Alexandra Munroe. New York: Harry N. Abrams, 2000.

Munroe, A., ed. *Yes Yoko Ono*. New York: Harry N. Abrams, 2000.

Nyman, M. *Experimental Music: Cage and Beyond*. Cambridge: Cambridge University Press, 1999.

Nyman, M. "Nam June Paik, Composer." In *Nam June Paik*, ed. John G. Hanhardt, 79–90. New York: Whitney Museum of Art in association with W. W. Norton, 1982.

Oschatz, S. Interview for *Frequency.* http://the.meso.net/oschatz/oval/jeremy.html.

Otomo, Y. Interview by Michel Henritz. *Revue et Corrigée*. English version can be found at http://www.japanimprov.com/yotomo/interview01.html.

Otomo, Y. "Leaving the Jazz Café." *Resonance* 4, no. 2 (1996): 4–7.

Owen, T. Review of *Faßt* (Touch, TO:32). *The Wire* 162 (August 1997): 52.

Papastergiadis, N. "'Everything That Surrounds': Art, Politics, and Theories of the Everyday." In *Everyday: 11th Biennale of Sydney*, 21–27. Sydney: Biennale of Sydney, 1998.

Parsons, M. "The Scratch Orchestra and Visual Arts." *Leonardo Music Journal* 11 (2001): 5–11.

Pisaro, K. "Music From Scratch: Cornelius Cardew, Experimental Music and the Scratch Orchestra in Britain in the 1960s and 1970s." Ph.D. diss., Northwestern University, 2001.

Pohlmann, K. *Principles of Digital Audio*. 4th ed. Madison, Wisconsin: McGraw-Hill, 2000.

Pohlmann, K. *The Compact Disc Handbook*. 2nd ed. Madison, Wisconsin: A-R Editions, 1992.

Popp, M. Interview with Heath Hignight. http://www.inaudible.com/artists/cross-selling/popp.html (no longer online).

Popp, M. Interview with Marc Weidenbaum. http://www.dnai.com/~marc/popp-script.html (no longer online).

Popp, M. Interview with *Splendid* magazine. http://www.splendidezine.com/features/oval/.

Portis, B., ed. *Christian Marclay: Cinema*. Oakville, Ontario: Oakville Galleries, 2000.

Poschardt, U. *DJ Culture*. Trans. Shaun Whiteside. London: Quartet Books, 1998.

Pritchett, J. *The Music of John Cage*. Cambridge: Cambridge University Press, 1993.

Project Dark, *Singles Club Catalogue.* http://www.projectdark.demon.co.uk/catalog.htm.

Purcell, L. *CD-R/DVD: Disc Recording Demystified*. New York: McGraw-Hill, 2000.

Ranada, D. "Digital Debut: First Impressions of the Compact Disc System." *Stereo Review* 48 (December 1982): 61–70.

Ranada, D. "Hands on the Digital Disc." *Stereo Review* 48 (April 1983): 61–63.

Ranada, D. "How to Buy a CD Player." *Stereo Review* 48 (July 1983).

Ranada, D. "Understanding Compact Disc Errors." *Stereo Review* 49 (April 1984): 48–50.

Revill, D. *The Roaring Silence: John Cage: A Life*. New York: Arcade Publishing, 1992.

Reynolds, S. *Energy Flash*. London: Picador, 1998.

Reynolds, S. "Low End Theories." *The Wire* 146 (April 1996).

Roberts, D. *All cracked Medias*. CD. Mille Plateaux, 1998.

Ronte, D. "Nam June Paik's Early Works in Vienna." In *Nam June Paik*, ed. John Hanhardt, 73–78. New York: Whitney Museum of American Art, 1982.

Ruhé, H. *Fluxus: The Most Radical and Experimental Art Movement of the Sixities*. Amersterdam: A, 1979.

Russolo, L. *The Art of Noises*. New York: Pendragon Press, 1986. http://www.ubu.com/papers/russolo.html.

Salter, L. "Compact Disc Reviews." *Gramophone* (March 1983).

Saltz, J. "Beautiful Dreamer." *Arts Magazine* 64 (December 1989).

Sangild, T. "The Aesthetics of Noise." http://www.datanom.com/noise/the_essay/index.asp.

Sasaki, A. "Dissolving Ground Zero: From 'Performance' to 'Sound.'" Liner notes for Ground Zero, *Last Concert*. CD. Amoebic, 1999.

Sasaki, A. "The Oscillating 'Will' and the Flickering 'Self,'" Liner notes to Filament. *29092000*. Trans. Yoshiyuki Suzuki and Cathy Fishman. CD. SAT/Amoebic, 2001.

Schütze, P. Review of Oval's *94 Diskont*. *The Wire* 135 (May 1995): 48.

Schwartz, J. "Jacques Attali's Revision of Baudrillard in 'Noise.'" *Art-OMMA* no. 9 (2003). http://www.art-omma.org/NEW/past_issues/theory/09.htm.

Serres, M. *Genesis*. Michigan: University of Michigan Press, 1995.

Serres, M. *The Parasite*. Baltimore: Johns Hopkins University Press, 1982.

Sherburne, P. "Thumbnail Music: Six Artists Talk About Minimalism." *Urbansounds*. http://www.urbansounds.com/us_current/thumbnail/ (no longer online).

Shiomi, M. "Meiko Shiomi." *Art and Artists* (October 1973).

Sinclair, I. *Audio and Hi-Fi Handbook*. Oxford: Newnes, 1998.

Smith, N. "The Splinter in Your Ear: Noise as the Semblance of Critique." *Culture, Theory, and Critique* 46: 1 (2005): 43–59.

Smith, S. "Compositional Strategies of the Hip-hop Turntablist." *Organised Sound* 5, no. 2 (2000): 75–79.

Sterne, J. *The Audible Past: Cultural Origins of Sound Reproduction*. Durham: Duke University Press, 2003.

Stiles, K. "Art and Technology." In *Theories and Documents of Contemporary Art*, ed. Kristine Stiles, 384–397. Berkeley: University of California Press, 1996.

Stiles, K. "Between Water and Stone." In *In the Spirit of Fluxus*, ed. J. Jenkins. Minneapolis: Walker Art Center, 1993.

Stiles, K. "Sticks and Stones: The Destruction in Art Symposium." *Arts Magazine* (January 1989): 54–60.

Stiles, K. "The Destruction in Art Symposium (DIAS): The Radical Cultural Project of Event-Structured Live Art." Ph.D. diss., University of California, Berkeley, 1987.

Stiles, K. "Thresholds of Control: Destruction Art and Terminal Culture." In *Ars Electronica: Facing the Future: A Survey of Two Decades*, ed. Timothy Druckrey, 124–134. Cambridge, Mass.: MIT Press, 1999.

Stiles, K. "Uncorrupted Joy: International Art Actions." In *Out of Actions: Between Performance and the Object, 1949–1979*, 226–238. Los Angeles: The Museum of Contemporary Art, 1998.

Stooss, T., and T. Kellein, eds. *Nam June Paik: Video Time—Video Space*. New York: Harry N. Abrams, 1993.

Stuart, C. "Damaged Sound: Glitching and Skipping Compact Discs in the Audio of Yasunao Tone, Nicolas Collins, and Oval." *Leonardo Music Journal* 13 (2003): 47–52.

Stuart, C. "The Object of Performance: Performativity in Contemporary Laptop Performance." *Contemporary Music Review* 22, no. 4 (2003): 59–65.

Stuart, C. "Yasunao Tone's Wounded and Skipping Compact Discs: From Improvisation and Indeterminate Composition to Glitching CDs." *Leonardo Electronic Almanac* 10: 9 (2002).

Sundall, M. “Jodi—New York—Internet Art of Joan Heemskerk and Dirk Paesmans Featured in INSTALL.EXE.” *Art Forum* (September 2003).

Szepanski, A. “Digital Music and Metatheory.” http://www.ritornell.net/theory/index.html (no longer online).

Tallman, S. “Always This Tüdeldität: Christian Marclay’s ‘Graffiti Composition.’” *Art On Paper* 4: 6 (2000): 28–33.

Tallman, S. “Sound and Vision: Multiples by Not Vital and Christian Marclay.” *Arts Magazine* 63 (May 1989): 17–18.

Taylor, C. “Noise Is OK.” *Semiotica* 52, no. 3/4 (1984).

Taylor, T. *Strange Sounds: Music, Technology, and Culture*. New York: Routledge, 2001.

Tetreault, M. “An Interview with Martin Tetreault.” By E. Baxter and P. Tagney. *Resonance* 5, no.1 (1996).

Thacker, E. “Bataille/Body/Noise: Notes towards a Techno-Erotics.” In *Merzbook: The Pleasure Dome of Noise*. Melbourne: Extreme, 1999.

Théberge, P. *Any Sound You Can Imagine: Making Music/Consuming Technology*. Middletown, Conn.: Wesleyan University Press, 1997.

Thomson, V. “John Cage Late and Early (1960).” Reprinted in *Writings about Cage*, ed. Richard Kostelanetz. Michigan: The University of Michigan Press, 1993.

Toch, E. “Über meine Kantate ‘Das Wasswe’ und meine Grammophonmusik.” *Melos* 9 (May/June 1930).

Tomii, R. “Concerning the Institution of Art: Conceptualism in Japan.” In *Global Conceptualism: Points of Origin, 1950s–1980s*, 15–29. New York: Queens Museum of Art, 1999.

Tomii, R. “Outsiders Looking In: Art Historiography by Non-Historians.” Paper given at College Art Association Annual Conference, New York, 2000.

Tomii, R. “State v. (Anti-)Art: Model 1,000-Yen Note Incident by Akasegawa Genpei and Company.” *Positions* 10, no. 1 (2002): 141–172.

Tomii, R. “Tokyo 1967–1973.” In *Century City: Art and Culture in the Modern Metropolis*, ed. Iwona Blazwick, 199–221. London. Tate Gallery Publishing, 2001.

Tomkins, C. “Video Visionary.” In *The Scene: Reports on Postmodern Living*, 195–226. New York: Viking Press, 5 May, 1976.

Tone, Y. “John Cage and Record.” *Intercommunication* 35 (2000).

Tone, Y. “John Cage and Recording.” *Leonardo Music Journal* 13 (2003): 11–15.

Tone, Y. Liner notes to *Solo for Wounded CD*. Tzadik, 1997.

Tone, Y. Transcript of an interview published in *Revue et Corrigée* 46 (2001).

Tone, Y., H. Naoyoshi, and A. Yukio, eds. “Chronology: Five Decades of Contemporary Art, 1916–1968.” *Bijutsu techõ* (April and May 1972).

Toop, D. “The Generation Game.” *The Wire* (May 2001): 39–44.

Toop, R. “Writing Music’s Boundaries.” http://www.ssla.soc.usyd.edu.au/conference/toop.html.

Underwood, M. “Communication Studies, Cultural Studies, Media Studies Infobase.” http://www.cultsock.ndirect.co.uk/.

Vogel, S. "In Record Time." *Artforum* 29 (May 1991).

Wilson, A. "Papa What Did You Do When the Nazis Built the Concentration Camps? My Dear, They Never Told Us Anything." In *Gustav Metzger: "Damaged Nature, Auto-Destructive Art."* London: Coracle and Lanmet, 1996.

Whitelaw, M. "Inframedia Audio." *Artlink* 21, no. 3 (2001): 49–52.

Wörner, F. H. *Stockhausen: Life and Work.* Trans. Bill Hopkins. London: Faber and Faber, 1973.

Yoshimoto, M. *Into Performance: Japanese Women Artists in New York*. New Brunswick: Rutgers University Press, 2005.

Young, L. M. "Lecture 1960." Reprinted in *Happenings and Other Acts*, ed. M. R. Sandford. London: Routledge, 1995.

Young, R "Worship the Glitch." *The Wire* 190/191 (December/January, 1999–2000): 52–57.

Zielinski, S. "Media Archaeology." http://www.ctheory.net/articles.aspx?id=42.

Zuberi, N. *Sounds English: Transnational Popular Music*. Urbana: Illinois University Press, 2001.

index